Plant Processing of Natural Gas

DOUG ELLIOT

J. C. KUO

PERVAIZ NASIR

published by

THE UNIVERSITY OF TEXAS
CONTINUING EDUCATION
PETROLEUM EXTENSION SERVICE
Austin, Texas

2008

Library of Congress Cataloging-in-Publication Data

Elliot, Doug, 1941-
  Plant processing of natural gas. — 2nd ed. / by Doug Elliot, J. C.
Kuo, [and] Pervaiz Nasir
      p.   cm.
  Includes bibliographical references and index.
  ISBN 978-0-88698-223-2 (alk. paper)
  1. Natural gas. 2. Gas manufacture and works. I. Kuo, J. C., 1950-
II. Nasir, Pervaiz, 1951- III. Title.
TP350.P57 2008
665.7'3—dc22
                          2007020068   CIP

Catalog No. 3.11020
ISBN 0-88698-223-2

*The University of Texas at Austin is an equal opportunity institution.
No state tax funds were used to print or mail this publication.*

Table of Contents

FIGURES   vi

TABLES   viii

FOREWORD   ix

ACKNOWLEDGMENTS   xi

ABOUT THE AUTHORS   xiii

CHAPTER 1.   Fundamentals   1
             Fluid Properties   1
                 Temperature   2
                 Pressure   3
                 Gravity and Miscibility   3
                 Solubility   4
             The Ideal Gas Law   4
                 Liquid Phase   5
                 Vapor Pressure   5
                 Boiling Point and Freezing Point   6
             Hydrates   7
             Comparing Physical Properties   8
             Composition   10
             Heat Energy   10
                 Heating Value   12
             Combustion   12
                 Flammability   13
             Applications   13
                 Flow Diagrams   14
             References   21

CHAPTER 2.   Feed Gas Receiving and Condensate Stabilization   23
             Treating and Processing   23
             Design Basis and Specifications for Treatment Units   26
                 Feed Gas Basis   26
                 Product Specifications   27
             Equipment Selection and Design   28
                 Pig Receivers   28
                 Slug Catchers   30
                 Condensate Stabilizers   32
                 Condensate Stabilizer Reboilers   32
                 Stabilizer Overhead Compressors   32
                 Gas and Liquid Heaters   32
             References   33

CHAPTER 3.   Dew-Point Control and Refrigeration Systems   35
             Process Descriptions   35
             Cost Estimate   35
                 Silica Gel Process   36
                 Glycol/Propane System   37
                 Glycol/J-T Valve Cooling Process   38
                 Comparison of Dew-Point Processes   40
                 Unit Specifications   40
             The Refrigeration System   41
                 Economizers   42
                 Chillers   44
                 Possible Problems   44
                 Multiple-Stage Refrigeration   46
             References   50

CHAPTER 4. Hydrocarbon Treating   51
            Gas-Treating Processes   51
            Chemical Reaction   51
                Amine-Based Solvents   52
                Nonamine-Based Processes   57
            Physical Absorption Processes   58
                Selexol®   59
                Propylene Carbonate Process   59
                Rectisol® Process   60
            Mixed Chemical/Physical Absorption   60
                Sulfinol® Process   60
            Adsorption on a Solid   61
                Molecular Sieve Process   61
                Activated Carbon Process   62
            Membrane Processes   62
            General Operating Considerations for Gas Treating   65
                Inlet Separation   65
                Foaming   65
                Filtration   65
                Corrosion   65
            References   66
CHAPTER 5. Sulfur Recovery and Claus Off-Gas Treating   67
            Sulfur Recovery   67
                Thermal Process   67
                Catalytic Recovery   68
            Claus Off-Gas Treating   70
                SCOT Process   70
            References   72
CHAPTER 6. Dehydration and Mercury Removal   73
            Dehydration   73
                Inhibitor Injection   76
            Dehydration Methods   80
                Liquid Desiccants   80
                Solid Desiccants   84
                Design Issues   88
            Mercury Removal Unit (MRU)   90
                Design Basis and Specifications   92
                Design Considerations   94
                Equipment Selection and Design   95
                Case Study   95
            References   96
CHAPTER 7. NGL Recovery—Lean-Oil Absorption   97
            Lean-Oil Absorption   98
                The Recovery System   98
                Absorption   99
                Why Absorbers Work   99
                Presaturation   100
                Potential Problems   102
            The Rejection System   104
                Hot Rich-Oil Flash Tank   104
                Rich-Oil Demethanizer   105
                Possible Problems   107
            The Separation System   108
                The Still   108
                Oil Purification   109
                Possible Problems   110
            References   111

CHAPTER 8.   NGL Recovery—Cryogenic   113
             Typical Applications   115
                 Turboexpander Process   115
                 Propane-Recovery Process   120
                 Ethane-Recovery Process   121
             Turboexpanders   122
             Cyrogenics   127
             References   130

CHAPTER 9.   Fractionation and Liquid Treating   131
             Fractionation   131
             Packed Columns   134
             NGL Fractionation Plants   134
                 Deethanizer (DeC$_2$) Column   136
                 Depropanizer (DeC$_3$) Column   137
                 Debutanizer (DeC$_4$) Column   137
                 Deisobutanizer (DIB) or Butane Splitter Column   137
             Product Specifications   138
             Monitoring Fractionation Plant Operation   139
             Possible Operating Problems   140
             NGL Product Treating   141
                 Liquid—Liquid Treating   141
                 Liquid—Solid Treating   143
             References   143

CHAPTER 10. Nitrogen Rejection Unit (NRU)   145
             Nitrogen Rejection   145
             NRU Process Selection   145
                 Pressure Swing Adsorption (PSA)   145
                 Cryogenic Absorption   145
                 Membranes   146
                 Cryogenic Distillation   146
             Cryogenic NRU Processes   146
                 Pretreatment   147
                 Chilling   148
                 Cryogenic Distillation   148
                 Recompression   148
             NRU Processes   149
             References   152

APPENDIX.   Figure and Table Credits   153
GLOSSARY   159
INDEX   183

Figures

1.1   Fluid molecules can be compared to marbles in a glass jar.   2
1.2   Comparison of different temperature measurement scales   2
1.3   Molecules escape and return to the liquid phase in a closed vessel.   5
1.4   Vapor pressure depends on the temperature of the liquid in a closed vessel.   5
1.5   Vapor pressure compared to the boiling temperature of liquid in a closed vessel   6
1.6   Hydrate plug in pipe   7
1.7   Vapor pressures of various hydrocarbons   9
1.8   Principles of sensible and latent heat   11
1.9   Light molecules vaporize and heavy molecules concentrate in a liquid.   14
1.10   Flow diagram symbols   15
1.11   Simple separation system for light and heavy fluid components   16
1.12   Simple separation system with reboiler   17
1.13   Simple separation system with reboiler and condenser   17
1.14   A series separation system   18
1.15   Bubble-cap tray for separation tower   19
1.16   Bubble-cap trays in tower   19
1.17   A tower separation system diagram   20
2.1   Gas processing plant   23
2.2   The functional units inside a gas processing plant   24
2.3   A typical feed gas receiving system   24
2.4   Diagram of a typical condensate stabilization system   25
2.5   Various types of pigs used for pipeline cleaning   29
2.6   A typical pig receiver system   29
2.7   Pig launcher/receiver   29
2.8   Finger-type slug catcher   30
2.9   Slug catcher being transported   31
2.10   Vertical slug catcher vessel   31
2.11   Horizontal slug catcher vessel   31
2.12   Pipe-fitting-type slug catcher   31
3.1   Diagram of a silica gel process   36
3.2   Glycol and propane refrigeration process   37
3.3   Glycol/J-T valve cooling process   39
3.4   A typical plant refrigeration system   41
3.5   Two different layouts of simple-staged separation process   42
3.6   Diagram of a refrigeration system using staged separation   43
3.7   Gas turbine-driven propane refrigeration compressor in a natural gas plant   43
3.8   Chiller diagram   44
3.9   Graph used to determine amount of ethane in propane   45
3.10   Graph used to determine amount of butane in propane   45
3.11   A two-stage refrigeration system   46
3.12   A three-stage refrigeration system   47
3.13   Diagram of a cascade refrigeration system with ethane and propane   48
3.14   Two-stage propane refrigeration system   49
4.1   Diagram of a typical aqueous amine treating plant   52
4.2   Schematic diagram of the Shell-Paques™ process   57
4.3   XTO Shell-Paques™ gas treating plant in Texas   58
4.4   Diagram of a typical physical solvent process   59

4.5	An integrated natural gas desulfurization plant	61
4.6	UOP's spiral wound membrane element	62
4.7	Element being inserted into the casing	63
4.8	A hollow-fiber membrane element	63
4.9	A membrane skid for the removal of CO_2 from natural gas	64
5.1	Species of elemental S in equilibrium at different temperatures	68
5.2	Three-stage modified Claus sulfur-recovery unit	69
5.3	A small, package-type, two-stage Claus plant that sends tail gas to an incinerator	69
5.4	Diagram of the Shell Claus Off-Gas Treating process (SCOT)	71
5.5	SCOT process plant	71
6.1	Water content of natural gas varies.	74
6.2	Pressure-temperature curves for predicting hydrate formation	75
6.3	Physical properties of selected glycols and methanol	76
6.4	A typical EG injection system	77
6.5	Freezing temperatures of ethylene glycol-water mixtures	78
6.6	Glycol reboiler temperatures	78
6.7	Hydrate depression versus minimum withdrawal concentration of ethylene glycol	79
6.8a	TEG dehydration unit	81
6.8b	Glycol regeneration unit designed to regenerate TEG for natural gas dehydration on an offshore oil and gas production platform	81
6.9	Effect of stripping gas on TEG concentration	82
6.10	Glycol regeneration processes	82
6.11	Equilibrium water dew points with various concentrations of TEG	83
6.12	TEG reboiler temperatures	84
6.13	Typical desiccant properties	85
6.14	Dry-bed dehydration unit schematic	86
6.15	Diagram of an adsorption tower	87
6.16	Mass transfer zone for water solid bed adsorption scheme	87
6.17	Horizontal filter separator	89
6.18	Mercury removal flow diagram	91
7.1	Oil absorption plant systems	98
7.2	Lean-oil-to-inlet-gas ratio	100
7.3	A typical low-pressure presaturation system using vapors from rich-oil demethanizer (ROD)	101
7.4	Accumulator	101
7.5	A residue gas scrubber diagram	102
7.6	Differential pressure indicators	103
7.7	Hot rich-oil flash tank used for methane rejection	104
7.8	Diagram of a ROD	105
7.9	Graph of demethanizer bottom temperature versus lean-oil-to-product ratio	106
7.10	Typical bottom temperature adjustments	107
7.11	The ROD reboiler is heated with hot lean oil in the bottom of the still.	107
7.12	Diagram of a dry still	108
7.13	An oil-reclaiming system design	109
7.14	A distillation test graph showing lean-oil quality	110
7.15	Losses due to poor quality of lean-oil initial boiling point	111
8.1	Lean-oil absorption process and cryogenic process	113
8.2	Pressure and temperature to recover 60% ethane	114
8.3	Pressure-temperature diagram for the turboexpander process	116

8.4	Diagram of a plant using a turboexpander process	117
8.5	Methane-ethane binary	118
8.6	Schematic of gas plant processing	119
8.7	Deethanizer overhead recycle process	120
8.8	Residue gas recycle process	121
8.9	A 3.5 turboexpander-compressor used to process offshore gas from the Gulf of Mexico	122
8.10	Efficiency of turboexpansion cooling	123
8.11	A radial-reaction turbine showing nozzle blades	123
8.12	Turboexpander	124
8.13	Wheel shaft	125
8.14	Active magnetic bearings for typical ABM turboexpanders	126
8.15	Core of an aluminum plate	127
8.16	Corrugated fin flow patterns	128
8.17	Cores of plate-fin heat exchanger (PFHE)	128
8.18	Components of a brazed aluminum heat exchanger	129
9.1	Diagram of a fractionation column	132
9.2	Flow through vapor passages	133
9.3	Various types of random packing	134
9.4	Tower with various packing materials including structured packing	135
9.5	Example of a four-column fractionation plant	135
9.6	Example of a fractionation plant used to produce three products	136
9.7	Fractionation plant	139
9.8	Schematic of a regenerable caustic process	142
10.1	Simplified block flow schematic for a nitrogen rejection facility	146
10.2	A single-column distillation unit	149
10.3	Double-column distillation process	150
10.4	Prefractionation scheme	151
10.5	Typical natural gas nitrogen phase	152

Tables

1.1	Physical Properties of Hydrocarbons Involved in Gas Processing	8
1.2	Physical Properties of Other Compounds Used in Gas Processing	9
1.3	Calculating Mole Percent	10
2.1	Typical Gas Feeds	27
4.1	Approximate Guidelines for Several Commercial Gas Processes	55
4.2	Physical Properties of Gas Treating Chemicals	56
9.1	Product Specifications for a Southern Louisiana Fractionation Plant	138
10.1	Components with Allowable Concentrations in a Cryogenic N_2 Rejection Process	147

Foreword

◆◆ The intent of the first edition of *Plant Processing of Natural Gas*, published in 1978, was to present the fundamental concepts and working practices involved in processing natural gas. The book was prepared primarily to aid plant employees who deal regularly with the problems of processing and treating gas.

The first edition was essentially written by a review committee appointed to ensure the book included positive, useful information, while minimizing highly technical material that would not contribute to the productive capability of plant employees.

The objective of this second edition is the same. However, the coauthors' goal in this book is to update old processing schemes and describe the state of the art that represents the evolution in plant processes and equipment. For example, this book focuses on cryogenic liquefaction versus the older oil absorption processes.

As before, the manual does not cover all materials, practices, procedures, or equipment used in the processing and treating of natural gas. Also, this book should not be interpreted as indicating a preference by the coauthors for one specific manner of operation. Readers must be guided by their own experiences in dealing with variances in local conditions and requirements.

It has been necessary and desirable to include photographs and drawings describing equipment that might be easily identified by manufacturer. The coauthors do not intend by these inclusions to infer in any way a preference or practice that is limited to equipment manufactured by any specific company.

Doug Elliot
PETEX Advisory Board
2008

Acknowledgments

❖ Over its many years of publication, *Plant Processing of Natural Gas* has been widely accepted by the gas processing industry and has been one of PETEX's best selling books.

The original book was written by a review committee. Those who contributed to the 1978 edition included the Texas Education Agency Department of Occupational Education and Technology, the Gas Processors Association, and the American Petroleum Institute – Committee on Vocational Training.

I would like to acknowledge Frank Overton of Shell Oil Company (retired) who made important contributions to the first draft of this revised text. I would also like to thank Warren True of the *Oil and Gas Journal (OGJ)* for his permission to use material previously published by *OGJ*.

Many thanks to Tim Ryan of Cryostar USA for generously contributing information and images about turboexpanders. Thanks also to Hank Traylor of UOP for assisting with material and images about membranes. I am grateful to Carmen Falk and Cathy Widner, both of IPSI, LLC, and Pat Elliot for providing administrative and clerical support to the authors.

I am indebted to Ron Brunner of the Gas Processors Association for allowing us to use and redraw many images from the 11th and 12th editions of the *GPSA Engineering Data Book*. The GPSA (Gas Processors Suppliers Association) schematics and flow diagrams are integral to helping the reader understand the complexities of the gas processing industry.

A special thank you to Amanda Koss, graphic artist at PETEX, who worked diligently to ensure the accuracy of many images, and to Sherry Rodriguez, who gathered photo permissions from the many contributors to this book.

Julia Ruggeri
PETEX Editor

About the Authors

❖ Dr. Doug Elliot has more than 40 years experience in the oil and gas business, devoted to the design, technology development, and direction of industrial research. Doug is currently President, COO and cofounder (with Bechtel Corporation) of IPSI LLC, a company formed in 1986 to develop technology and provide conceptual design services to oil and gas producing and engineering, procurement, and construction companies.

Prior to IPSI, Doug was Vice President of Oil and Gas with Davy McKee International. Doug started his career with McDermott Hudson Engineering in the early 1970s following a postdoctoral research assignment under Professor Riki Kobayashi at Rice University, where he developed an interest in oil and gas thermophysical properties research and its application.

Doug has authored or coauthored over 65 technical publications plus 12 patents. He served as a member of the Gas Processors Association Research Steering Committee from 1972 to 2001 and as Chairman of the GPSA (Gas Processors Suppliers Association) Data Book Committee on Physical Properties. Doug served as Chairman of the South Texas Section and Director of the Fuels and Petrochemical Division of the American Institute of Chemical Engineers; and is currently a member of the PETEX Advisory Board. He holds a B.S. degree from Oregon State University and M.S. and Ph.D. degrees from the University of Houston, all in chemical engineering.

Doug is a Bechtel Fellow and a Fellow of the American Institute of Chemical Engineers.

◆◆ J.C. Kuo (Chen Chuan J. Kuo) is a 34-year veteran of the gas processing, gas treating, and liquefied natural gas (LNG) industry. As a senior advisor for Chevron's Energy Technology Company, he has served as the Process Manager/ Process Lead for many projects, including the Wheatstone LNG, Gorgon LNG, Delta Caribe LNG, Casotte Landing, and Sabine Pass LNG terminal projects. He has also served as the technical process reviewer for Angola, Olokola, Algeria, and Stockman LNG projects. Before working at Chevron, J.C. was the Technology Manager for IPSI, an affiliate of Bechtel, and served as the Process Manager/Process Lead for the Pemex Catarell offshore project, the Egyptian LNG (Idku) trains 1 and 2, China Shell Nan Hai, Chevron Venice gas plant de-bottleneck, Tunisia NRU, and Australian SANTOS projects.

J.C. is a frequent speaker and presenter at international conferences such as for the American Institute of Chemical Engineers, gas processing and treating conferences, and the LNG Summit. He has contributed to gas processing and LNG technology improvements through a patent, a book, and many papers. He has also served as co-chair of the AIChE LNG sessions for the topical conferences on natural gas utilization. He is a member of the steering committee for the North American LNG Summit.

His degrees include a B.S. from Chung Yuan Christian University, Taiwan, and an M.S. from the University of Houston, both in Chemical Engineering; and an M.S. in Environmental Engineering from Southern Illinois University. He is a registered Professional Engineer in Texas and a member of AIChE. He is the president of the 99 Power Qi Qong Texas divisions.

❖ Dr. Pervaiz Nasir has more than 27 years experience in the oil and gas business. He is currently the Regional Manager Gas/Liquid Treating and Sulfur Processes, Americas, at Shell Global Solutions.

Pervaiz started his career at Shell Development Company in 1981 in research and development and technical support, mostly related to oil and gas processing. In 1986, he moved into licensing and process design of Shell Gas/Liquid Treating technologies. As a member of Shell Midstream from 1991 through 1999, Pervaiz was responsible for the operations support and optimization of existing gas plants and the development and startup of new gas processing facilities. He then joined Enterprise Products Company as Director of Technology. At Enterprise, Pervaiz was responsible for the evaluation of new business ventures/technologies in gas processing, liquefied natrual gas (LNG), petrochemicals, etc. He returned to Shell Global Solutions in 2006.

Pervaiz holds a B.S. from Middle East Technical University (Ankara), an M.S. from University of Alberta (Edmonton), and a Ph.D. from Rice University (Houston), all in chemical engineering. He served on the Gas Processors Association (GPA) Phase Equilibria Research Steering Committee from 1983 through 1990 and is currently a member of the GPA LNG Committee. Pervaiz has authored or coauthored over 17 external technical publications.

Natural gas is colorless, shapeless, and odorless in its pure form. It is a fossil fuel consisting primarily of methane with quantities of ethane, propane, butane, pentane, carbon dioxide, nitrogen, helium, and *hydrogen sulfide*. Natural gas is combustible, gives off a great deal of energy, is clean burning, and emits low levels of byproducts into the air. It is an important source of consumer energy used in homes to generate electricity.

1
Fundamentals

The petroleum industry classifies natural gas by its relationship to crude oil in the underground *reservoir*. *Associated gas* is the term used for natural gas that is in contact with crude oil in the reservoir. The associated gas might be a *gas cap* over the crude oil in a reservoir or a solution of gas and oil. *Nonassociated gas* is found in a gas phase in reservoirs without crude oil.

Whether associated or nonassociated, *gas production streams* are highly variable and can contain a wide range of hydrocarbon and nonhydrocarbon components. These streams might include various mixtures of liquids and gases as well as solid materials. There are usually some nonhydrocarbon components including nitrogen, helium, carbon dioxide, hydrogen sulfide, and water vapor present in the stream. Trace amounts of other components, such as mercury, might also be present.

Natural gas processing plants use physical and chemical processes to separate and recover valuable *hydrocarbon* fluids from a gas stream. In the processing plant, all the pipes, containment vessels, steam lines, tanks, pumps, *compressors*, towers, and instruments contain a gas or liquid undergoing some kind of treatment process.

During the processing, the nonhydrocarbon *contaminants* must be handled properly because they affect gas behavior during treatment, impair the efficiency of processing operations, or can damage the processing equipment. For example, the contaminant, liquid mercury, weakens and bonds with the aluminum *heat exchangers* used to produce supercooled fluids. If mercury is not removed from the gas early in the processing phase, it liquefies and collects on the exchanger's surfaces, eventually destroying the heat exchangers.

FLUID PROPERTIES

When there is a pipe, a steam line, a tank, a pump, a compressor, a tower, an instrument, or even a filled sample container in a gas plant, it almost always contains a *fluid*.

What is a fluid? A fluid can be a gas, a liquid, or a solid. A fluid is defined as any substance that flows freely unless restricted or contained by a barrier. Without the ability to assume a shape of its own, a fluid assumes the shape of the container into which it is placed. Both gases and liquids are classified as fluids.

Natural *gas treatment* is based on the reactions of reservoir fluids in physical and chemical processes. Each fluid has a unique set of properties including gravity, solubility, and flammability controlling its response to given stimuli. A gas processing plant operator must determine the specific properties and conditions of its source of oilfield fluids, or *feedstock,* because each one is different. Problems that occur during gas processing come from a fundamental misunderstanding of the specific fluid properties or the physical and chemical laws that determine fluid behavior.

Figure 1.1 Fluid molecules can be compared to marbles in a glass jar.

To understand gas processing, it is important to understand the fundamental principle of fluid composition. At the simplest level, everything is composed of atoms combined into groups called molecules. How fluids behave depends on how their molecules behave.

Imagine a glass jar filled with marbles in which each marble represents a *molecule* and all the marbles together represent a fluid (fig. 1.1). Because fluids are made of molecules, the behavior of oilfield fluids is determined by their molecular behavior. *Process engineers* use molecular behavior in design calculations and plans for gas processing plants.

To understand gas plant processing, it is important to become familiar with the following key terms and concepts used throughout this book.

Temperature

Manipulating temperature during gas processing is essential in separating the gas into marketable products. In the gas processing industry, *temperature* is usually expressed in degrees *Fahrenheit* (F) or degrees *Rankine* (R). Zero degrees F is the temperature at which water freezes. Zero degrees R equals *absolute zero* (about –460°F) and is the theoretical point at which all molecular motion stops. To convert degrees F into degrees R, simply add the number 460 to the F value. Therefore, 100°F is equal to 560°R.

Engineering calculations that deal with gas behavior, such as compressor cylinder capacity and compressor discharge temperatures, use absolute zero temperatures. The Fahrenheit and Rankine scales are illustrated in figure 1.2, along with the centigrade and *Kelvin* temperature scales.

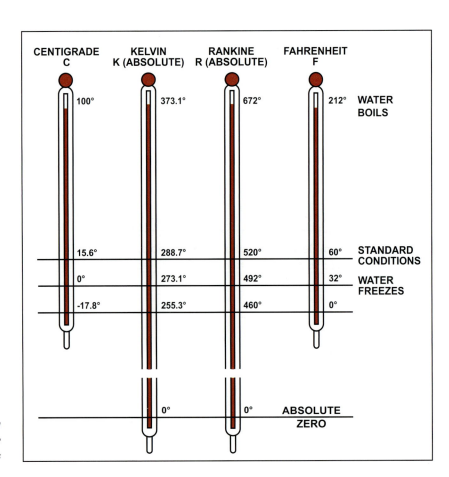

Figure 1.2 Comparison of different temperature measurement scales

Critical temperature is the highest temperature at which a substance can be separated into two distinct fluid phases—liquid and vapor. Above the critical temperature, a gas cannot be liquefied by pressure alone. The critical temperature of each substance in natural gas varies and must be understood to carry out the processing safely and effectively.

Pressure

Pressure is the force that a fluid, either liquid or gas, applies uniformly in all directions within a vessel, pipe, or other contained space. Pressure is measured in terms of pounds of force exerted within a one-square-inch area known as *pounds per square inch (psi)*. The pressure required to condense a vapor at critical temperature is referred to as the vapor's *critical pressure*. In gas processing, specific liquids cannot be separated if the pressure is greater than the critical pressure of either the mixture or the separated fluids.

A fluid exerts a certain amount of pressure in the interior of the vessel that contains it. This is measured with a pressure gauge and is known as *gauge pressure*. Gauge pressure is a useful expression of the difference between the pressure inside the container and the atmospheric (external) pressure. The gauged pressure of a fluid inside a containment vessel is measured in *pounds per square inch gauge (psig)*.

Absolute pressure is the true pressure of a contained fluid. It is the sum of psig and *atmospheric pressure* and written as *pounds per square inch absolute (psia)*. For example, atmospheric pressure at sea level is 14.7 psi. Therefore, the absolute pressure in a tank at sea level is the gauge pressure (psig) plus 14.7 psi.

Gravity and Miscibility

Gravity is an expression of fluid weight. The gravity of hydrocarbon liquids, like *propane*, gasoline, or crude oil, is expressed in degrees of *API gravity*. API (American Petroleum Institute) gravity is the measure of petroleum liquid density compared to the density of water. The lighter the petroleum liquid, the higher its API gravity will be.

Specific gravity relates the density or weight per unit *volume* of one fluid to the density of another fluid at the same temperature. The density of a liquid is usually compared to that of water. The density of a gas is compared to that of air.

When two liquids do not or cannot mix together, they are called *immiscible*. If combined, the two fluids separate with the lighter liquid floating to the top. For example, oil and water are immiscible, and oil is lighter than water. When placed in a tank, the two liquids will separate, and most of the oil will float to the top. At this point, almost all of the water can be drained from the bottom of the tank. This water is called *free water* because it can be easily separated from a gas or liquid.

Miscible liquids are liquids with different specific gravities that can be mixed together without separating. For example, *butane* is lighter than gasoline, but if the two liquids are put into the same tank, they will not separate. All hydrocarbons in a gas processing plant are miscible with one another but are immiscible with fluids such as water and glycol.

Solubility

The *solubility* of a gas or liquid, is the degree to which a substance will dissolve in a particular solvent. Solubility is an important consideration in gas processing when mixing hydrocarbons with fluids such as *glycol*. Glycol is a compound often used in the *dehydration* or water removal processes during gas treatment. Because glycol and hydrocarbons are slightly soluble, a major gas processing concern is that glycol can be contaminated and degraded by hydrocarbons.

Gasoline and water are considered immiscible. Although immiscible, water and gasoline are *mutually soluble*. A small quantity of water will dissolve in the gasoline, and some gasoline will dissolve in the water. Less than a half pound of water will dissolve in 1,000 pounds of gasoline. However, at low temperatures, even a small amount of water can freeze and cause problems in gas processing.

THE IDEAL GAS LAW

Fluids can exist in a gas, liquid, or solid state, but gas processing deals only with fluids in gas and liquid phases. Gases are sometimes called *vapors*, and the terms are technically interchangeable.

A gas or vapor is composed of millions of rapidly moving molecules colliding with each other and anything that contains them. Pressure is caused by the force of the moving gas molecules colliding with the walls of the containment vessel. Altering either the temperature or volume of gas in a closed vessel will change the pressure. Decreasing the number of collisions or distributing the gas over a larger area will reduce the gas pressure. Increasing the number of collisions or reducing the area of collisions will increase the gas pressure.

Heated gas makes the molecules move faster; they strike each other and the vessel walls more often, and pressure in the vessel increases. When gas is cooled, its molecules move more slowly and the pressure decreases. If a volume of gas expands into a larger vessel, the distance between the gas molecules increases. The molecules do not hit each other or the vessel's walls as often, so the pressure of the gas decreases. Conversely, compressing a gas forces its molecules into a smaller volume. The collisions of the gas molecules are concentrated into a smaller area, so the gas pressure increases.

The *ideal gas law* states that all gases react in the same way. The ideal gas law can be written as the following equation:

$$Pv = RT$$

or

$$PV = nRT$$

where

R = a proportionality factor
T = absolute temperature
v = volume of one mol of gas
n = number of mols of gas
V = volume of n mols of gas
P = absolute pressure

If no liquid is present, all ideal gases have the same pressure if contained in vessels of the same size and held at the same temperature. Another way to state the ideal gas law is that all gases occupy an equal volume at the same conditions of temperature and pressure.

Liquid Phase

The molecules of a liquid are held together by a natural attraction that keeps each molecule confined to a small space. Within that space, however, a molecule can be moving violently.

In a partially filled vessel, violent molecular motion at the surface of a liquid can cause molecules to escape the fluid. The escaped molecules form a vapor above the liquid. However, this vapor can also revert back to a liquid phase. As the newly formed gas molecules move toward the surface of the liquid, those hitting the liquid tend to stick. At the same time that some molecules change to the gas phase, others are returning to the liquid phase. The process is shown in figure 1.3.

Vapor Pressure

When the number of molecules leaving the liquid equals the number of molecules entering the liquid, the gas and liquid are said to be in *equilibrium*. Equilibrium means that the number of molecules moving from gas to liquid and from liquid to gas is equal. It does not mean the gas volume is equal to the liquid volume.

At equilibrium conditions, the pressure of the gas phase is equal to the vapor pressure of the liquid phase. If the temperature of liquid in a closed vessel increases, the vapor pressure increases until the vapor and liquid are at equilibrium (fig. 1.4).

When more molecules escape from the liquid than are returning, the liquid is *vaporizing* or becoming a vapor. When more molecules are entering the liquid from the vapor than are escaping from the liquid, the vapor is *condensing*. The phase changes from vapor to liquid or liquid to vapor are caused by a change in the distance between molecules. The molecules of a gas are farther apart than the molecules of a liquid. A change in phase does not change the molecular structure or chemical nature of the fluid.

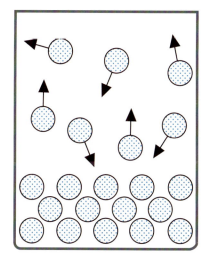

Figure 1.3 Molecules escape and return to the liquid phase in a closed vessel.

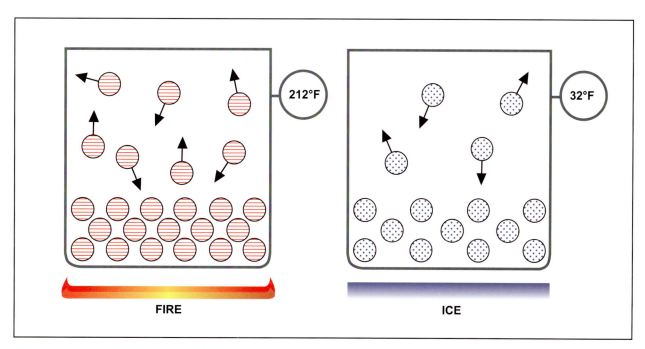

Figure 1.4 Vapor pressure depends on the temperature of the liquid in a closed vessel.

When a liquid is heated and free to expand, it expands in all directions. This is why *liquid expansion* is also known as *volume expansion*. Often, gas operators bottle a sample of gasoline when it is cold, only to have the cork pop off the bottle when the gasoline warms to room temperature.

If a pressure vessel is completely filled, volume expansion can cause rupturing. The liquid under pressure can also immediately vaporize and cause an explosion. To prevent volume expansion in gas processing, vapors are vented to a vapor recovery system. The *Gas Processors Suppliers Association (GPSA)* recommends that when filling pressure containers at temperatures above 0°F, operators should leave at least 20% of the container empty to allow for the volume expansion that will occur when the sample is brought into a warm room. The U.S. Department of Transportation and the National Surface Transportation Board have similar specifications for loading railroad tank cars with a liquid petroleum gas such as butane.

Boiling Point and Freezing Point

A liquid boils when its vapor pressure becomes equal to the pressure in the vessel. To make a contained liquid boil when vessel pressure is fixed, the temperature must be increased. To stop the liquid from boiling, the temperature must be lowered.

For different liquids, at each pressure, there is a temperature at which that liquid will boil. This is known as the *boiling point* or boiling temperature (fig. 1.5). For each temperature there is a pressure—the vapor pressure of the liquid—at which the liquid will boil.

Pure hydrocarbon vapors and liquids can also freeze but only at very low temperatures. This is generally not a problem in gas processing plants. However, if free water is present in the vapors and liquids, it can freeze, even at temperatures above 32°F.

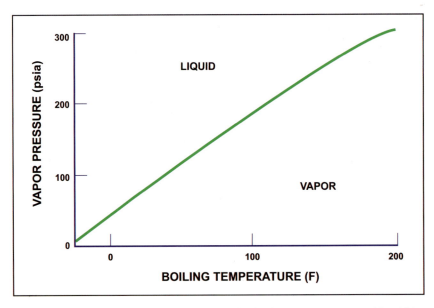

Figure 1.5 Vapor pressure compared to the boiling temperature of liquid in a closed vessel

HYDRATES

Free water in a gas plant can cause *hydrates* to form. Hydrates are an icy mixture of hydrocarbons and water that can form at temperatures as high as 80°F, depending on the pressure. Processing plant operators carefully monitor the operation to prevent hydrates from forming. Hydrates can plug lines, valves, heat exchanger tubes, and even vessels. *Hydrate plugs* can be controlled or eliminated by injecting *methanol* or glycol into the fluid, by increasing the temperature of the fluid, or by lowering the pressure (fig. 1.6).

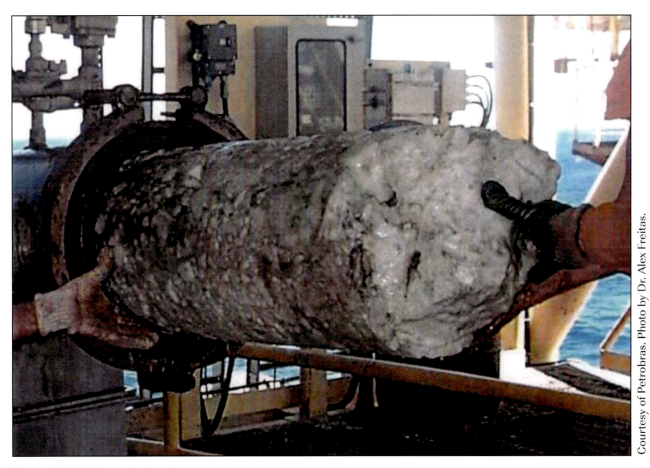

Courtesy of Petrobras. Photo by Dr. Alex Freitas.

Figure 1.6   Hydrate plug in pipe

COMPARING PHYSICAL PROPERTIES

Table 1.1 lists the primary gas processing hydrocarbons and some of their more important physical properties. The lightest hydrocarbons are at the top of the table and the heaviest at the bottom. An abbreviated chemical symbol for each is listed after the name. While these are not the designations chemists use, they are the abbreviations commonly used in gas processing plants.

The third column in Table 1.1 lists the specific gravity of each hydrocarbon in a liquid phase at 60°F. Specific gravity changes with temperature because liquids are lighter at higher temperatures. When stating the specific gravity of a liquid, the temperature must also be given.

The fourth column in Table 1.1 shows the specific gravity of each hydrocarbon in its gas phase. This is expressed as the weight of a specified volume of gas compared to the weight of the same volume of air. The specific gravity of air is 1.00. Only *methane* or a gas mixture with lots of methane is lighter than air. If a vessel full of methane is vented, the escaping vapor will rise. However, if a tank containing propane or any of the heavier hydrocarbon gases is vented, the vapor will settle to the ground and become a potential hazard.

The next two columns of Table 1.1 list the boiling point at 14.7 psia and the vapor pressure at 100°F. The boiling point of a fluid depends on its pressure.

Figure 1.7 charts the vapor pressure and boiling temperature for some typical hydrocarbons and other fluids. If the temperature of a fluid is known, this chart can be used to find its vapor pressure. For example, if a propane surge tank is at 100°F, then the vapor pressure of the propane is 190 psia.

Table 1.2 lists some of the physical properties of fluids that are often used in gas processing.

Table 1.1

Physical Properties of Hydrocarbons Involved in Gas Processing

Hydrocarbon	Symbol	Liquid Specific Gravity @ 60°F	Gas Specific Gravity	Boiling Point @ 14.7 psia °F	Vapor Pressure @ 100 °F psia	Critical Temp. °F	Critical Pressure psia
Methane	C_1	0.30	0.55	-259	5,000	-117	668
Ethane	C_2	0.36	1.04	-127	800	90	708
Propane	C_3	0.51	1.52	-44	190	206	616
Isobutane	iC_4	0.56	2.01	11	72	275	529
Normal Butane	nC_4	0.58	2.01	31	52	306	551
Isopentane	iC_5	0.62	2.49	82	20	369	490
Normal Pentane	nC_5	0.63	2.49	97	16	386	489
Hexane	C_6	0.66	2.97	156	5.0	454	437
Heptane	C_7	0.69	3.46	209	1.6	513	397
Octane	C_8	0.71	3.94	258	0.5	564	361
Nonane	C_9	0.72	4.43	303	0.2	611	332
Decane	C_{10}	0.73	4.91	345	0.1	652	304

Source: *GPSA Engineering Data Book*, 12th edition

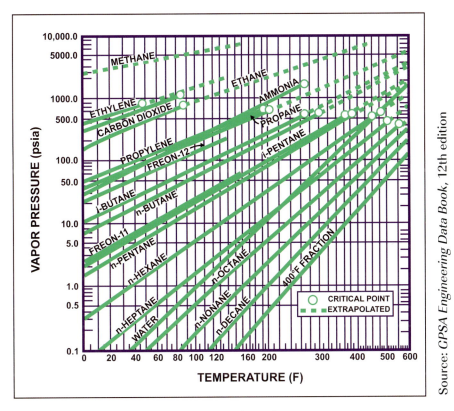

Figure 1.7   Vapor pressures of various hydrocarbons

Source: *GPSA Engineering Data Book, 12th edition*

Table 1.2

Physical Properties of Other Compounds Used in Gas Processing

Other Compounds	Liquid Specific Gravity @ 60°F	Boiling Point @ 14.7 psia °F	Vapor Pressure @ 100°F psia	Freezing Point @ 14.7 psia °F	Fire Point @ 14.7 psia °F
Nitrogen	0.81	-320	-	-	-
Carbon Dioxide	0.83	-109	1,250	-	-
Hydrogen Sulfide	0.79	-79	394	-	-
Water	1.00	212	0.9	32	-
Monoethanolamine (MEA)					
100%	1.02	339	-	51	200
15%	1.01	213	-	20	-
Triethylene Glycol (TEG)					
100%	1.13	546	-	19	330
Ethylene Glycol					
100%	1.11	387	-	8	245
70%	1.09	240	-	-80	-

Source: GPA publication 2145[27]

Fundamentals

COMPOSITION

A *mole* is the base unit used to express the mass or amount of a substance based upon its *molecular weight*. This weight can be expressed in grams, pounds, tons, or what unit is required. For example, the molecular weight of water is 18 because the *atomic weight* of each of water's two hydrogen atoms is 1, and the atomic weight of water's single oxygen atom is 16. One gram-mole of pure water weighs 18 grams; a pound-mole of pure water weighs 18 pounds. Similarly, one pound-mole of methane (CH_4) is 16 pounds because the molecular weight of methane is 16.

For mixtures of gases, the term *mole percent* (mol %) expresses the ratio of the number of moles of a constituent gas to the total number of moles in a mixture. Mole percent is often used to express the quantity of each gas in a mixture. Mole percent and *volume percent* are the same for any mixture of gases. However, mole percent is not equal to volume percent for a mixture of liquids.

Table 1.3 shows the mole percent of each component in a mixture of four gases. The number of moles of each gas is determined by dividing the actual weight of the gas by its molecular weight because a mole is numerically equal to the component's molecular weight.

Table 1.3
Calculating Mole Percent

Component	Symbol	Molecular Weight	Weight of Component	Calculated Number of Moles	Mole Percent
Methane	C_1	16	60 lb	$60/16 = 3.750$ moles	$3.750/4.816 \times 100 = 77.87\%$
Ethane	C_2	30	20 lb	$20/30 = 0.667$ mole	$0.667/4.816 \times 100 = 13.85\%$
Propane	C_3	44	10 lb	$10/44 = 0.227$ mole	$0.227/4.816 \times 100 = 4.71\%$
Butane	C_4	58	10 lb	$10/58 = 0.172$ mole	$0.172/4.816 \times 100 = 3.57\%$
Totals			**100 lb**	**4.816 moles**	**100.00%**

HEAT ENERGY

Matter exists in three phases: solid, liquid, and gas. The amount of *heat energy* in a substance determines the phase in which the substance exists. Adding or taking away enough heat energy can transform a substance into a different state of matter.

Adding heat to solid ice can transform it into liquid water. Adding heat to water can cause it to boil and transform into a gas or steam. Reducing heat energy in a substance can reverse the process, changing steam to water and water to ice as the temperatures drop.

Scientists measure heat in *British thermal units* (*Btus*). The term Btu describes a specific amount of thermal energy, just as the word *horsepower* describes a specific amount of mechanical energy. One Btu is defined as the amount of heat needed to increase the temperature of one pound of water by one degree Fahrenheit.

Heat that changes the temperature of a fluid without changing it into a gas or solid is known as *sensible heat*. The sensible heat capacity of a fluid is a measure of its ability to accept or release heat without changing its phase.

Heat that changes the phase of a fluid is called *latent heat*. If a boiling liquid is heated to make more vapor, the heat added is latent heat. If heat is taken from a condensing vapor to make more liquid, then latent heat is being removed. Sensible heat and latent heat are illustrated in figure 1.8.

One pound of water at 212°F contains less heat energy than one pound of steam at 212°F. The difference is the latent heat required to complete the phase change from water to steam.

The difference between sensible and latent heat depends upon the specific fluid involved, its phase, temperature, and pressure. Generally, it takes more than about 250 Btu of heat to make one pound of vapor from a given hydrocarbon liquid than to raise the temperature of one pound of the liquid by one degree Fahrenheit. Similarly, the hydrocarbon vapor can be condensed back into a liquid by removing 250 Btu of heat.

The difference between sensible and latent heat for water is greater. It takes 1,000 Btu more heat to make one pound of steam from water than it takes to raise the temperature of one pound of water by one degree Fahrenheit.

Whenever solids, liquids, or gases with different temperatures come into contact, heat is transferred between the two substances. If an object or substance feels warm to the touch, heat energy is transferring from the object or substance to the hand. Conversely, if an object or substance feels cold to the touch, heat is being transferred from hand to object.

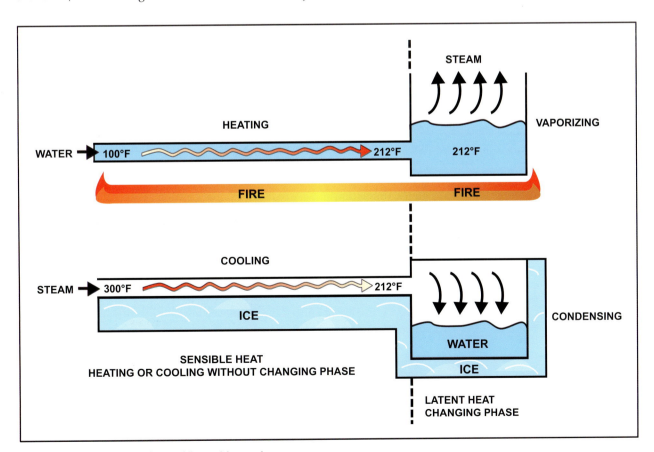

Figure 1.8 Principles of sensible and latent heat

According to the principles of heat transfer, a substance cannot be cooled below the temperature of its cooling medium or a temperature hotter than its heating medium. An amount of heating or cooling medium and temperature *differential* is required to drive the heat exchange in the desired direction. Water at 85°F can cool a fluid to 120°F from a temperature of 200°F if there is enough water and the water temperature remains lower than the fluid being cooled.

Heating Value

The amount of heat produced by the complete *combustion* of a unit of matter is referred to as its *heating value* or the heat of combustion. Methane, for example, has a heating value of about 900 Btu per *standard cubic foot (scf)*. When a natural gas composed of 90% methane and 10% nitrogen and carbon dioxide is burned, the heating value of the gas is somewhat less than 900 Btu per scf. This is because nitrogen and carbon dioxide have no heating value.

The heating value of natural gas in gas sale contracts is calculated either as *gross heating value* and *net heating value*. It is important to know the difference. Gross heating value is the sum of the normal heating value of methane plus the heating value of the other components in the gas.

The gas-combustion process produces water as a byproduct. The process of vaporizing water into steam consumes heat energy reducing the heating efficiency of the fuel. The net heating value of natural gas is the difference between its gross heating value and the heat energy, required to vaporize water *entrained*, or trapped, in the gas.

COMBUSTION

Combustion is a chemical reaction between oxygen and a combustible material. The three essential requirements for combustion are oxygen, fuel, and a source of ignition such as heat energy. Removing any one of these components will extinguish a fire or prevent it from starting.

Combustion is complete when there is enough oxygen to convert all of the carbon in the fuel to carbon dioxide and all of the hydrogen in the fuel to water. Anything less than complete combustion leaves unburned or partially chemically changed fuel in the exhaust.

In a gas processing plant, the two major causes of heat loss are incomplete combustion and *dry gas* loss. Changing the ratio of fuel-to-combustion air usually controls these losses. To determine how efficiently a fuel is burning, it is necessary to analyze the *off-gases* (or *flue gas*) from combustion to find the concentration of by-products formed by combustion. A flue gas containing 2.5% oxygen indicates that the fuel mixture contains about 15% excess air, which is the desired level for most fired heaters and boilers.

If the flue gas analysis shows an oxygen content below 2.5%, some of the fuel is not being burned. Incomplete combustion wastes fuel and creates unwanted emissions that must be removed from the flue gas.

If the flue gas contains more than 2.5% excess oxygen (more than 15% excess air), then some of the heat energy is disappearing up the exhaust stack with the excess air passing through the furnace. This is known as dry gas loss or stack loss. Dry gas loss caused by excess air increases fuel consumption. When the flue gas shows an oxygen content above 5% or the equivalent of 30% excess air, the excess air should be removed.

Flammability

Not all vapor and air mixtures are flammable. Every gas has a specific flammability range at which the fuel-to-air ratio will ignite. At atmospheric pressure, the low end for hydrocarbon gas processing is about 1% hydrocarbon and 99% air. At the high end, gas processing hydrocarbons will ignite with 15% hydrocarbon and 85% air.

Hydrocarbons will burn only if the air-to-fuel ratio is within the flammable range. Below 1% hydrocarbons, there is not enough fuel for ignition. Above 15% hydrocarbons, there is not enough oxygen.

Increasing the pressure widens the flammable range of a hydrocarbon and air mixture. If air has entered a closed vessel containing hydrocarbons, the mixture can be explosive. Safe operation might be recovered by decreasing the pressure within the vessel. Precautions must be taken to keep the potentially explosive vapor from making contact with a source of ignition.

The *flash point* of an ignited liquid is the temperature at which sufficient vapors are produced to cause a momentary flash but not enough to develop into a continuous burn. The lowest temperature at which a liquid produces enough vapors to burn continuously is called the *fire point*. If a liquid is burning, the burning gases are adding heat to the liquid. This extra heat keeps the liquid vaporizing fast enough to support combustion and keeps the liquid at its fire-point temperature.

APPLICATIONS

Nearly all gas processing operations are done on fluid mixtures composed of various individual hydrocarbons. Pure fluids are rare in gas processing operations. When a mixture's vapor and liquid are in equilibrium, the various components present in the liquid are also present in the vapor. Equilibrium is the reason that gas processing requires so many different operations. While separating desired hydrocarbons from the inlet gas stream, some undesirable components remain in the mix. Inevitably, removing unwanted components also removes some of the desired products.

A pure fluid has a specific boiling point at a given pressure. At the boiling-point temperature, a pure liquid will continue to boil until all is vaporized. A mixture of several fluids will have a wider range of boiling temperatures. If the temperature is increased until the mixture starts to boil at the *initial boiling point,* some gases in the mixture will begin to vaporize. In working with pure hydrocarbon fluids, the light liquids have low boiling-point temperatures and high vapor pressures. The heavy liquids have high boiling-point temperatures and low vapor pressures. Light liquids vaporize or are more volatile than heavy liquids. Light molecules escape the liquid more rapidly than heavy molecules in a mixture of hydrocarbon fluids.

Light molecules tend to concentrate in the vapor, and heavy molecules tend to concentrate in the liquid, as shown in figure 1.9.

The amount of light and heavy components remaining in a mixture is partially controlled by the temperature of the boiling liquid. When all the gases at the initial boiling point have been vaporized, boiling will then stop unless the temperature is increased. A gas mixture will not completely vaporize until the treatment temperature reaches the boiling point of each component gas.

Volatility is the term used to describe how easily a given fluid will vaporize. Differences in volatility make gas processing possible. Safe plant operations require that workers know the relative volatilities of each individual hydrocarbon in the gas stream.

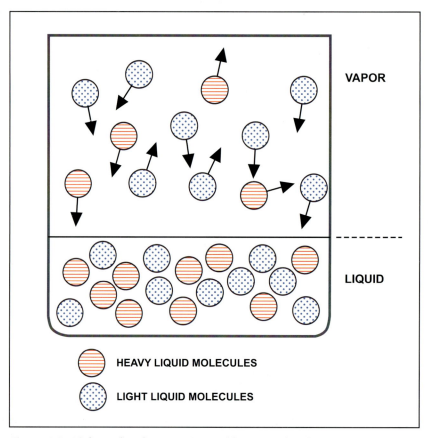

Figure 1.9   Light molecules vaporize and heavy molecules concentrate in a liquid.

Flow Diagrams

In a gas processing plant, several items of equipment function as a unit. Units are often illustrated in *flow diagrams* with symbols representing various types of equipment. Figure 1.10 shows some common symbols used in flow diagrams.

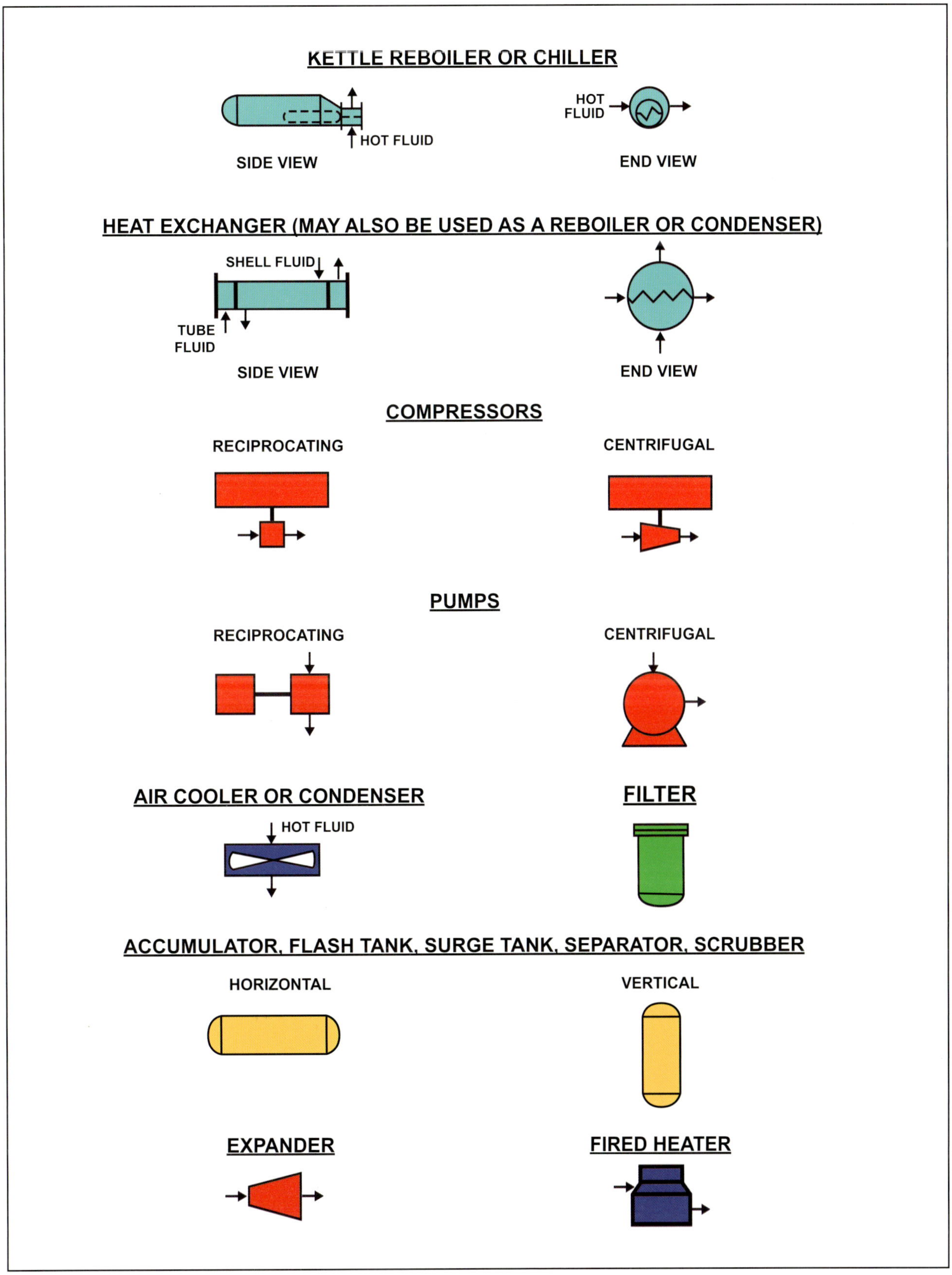

KETTLE REBOILER OR CHILLER

HOT
FLUID →

↑ HOT FLUID

SIDE VIEW   END VIEW

HEAT EXCHANGER (MAY ALSO BE USED AS A REBOILER OR CONDENSER)

SHELL FLUID ↓ ↑

TUBE ↑ ↓
FLUID

SIDE VIEW   END VIEW

COMPRESSORS

RECIPROCATING   CENTRIFUGAL

PUMPS

RECIPROCATING   CENTRIFUGAL

AIR COOLER OR CONDENSER   **FILTER**

↓ HOT FLUID

↓

ACCUMULATOR, FLASH TANK, SURGE TANK, SEPARATOR, SCRUBBER

HORIZONTAL   VERTICAL

EXPANDER   **FIRED HEATER**

Figure 1.10   Flow diagram symbols

Source: *GPSA Engineering Data Book*, 12th edition

Shown in figure 1.11 is a flow diagram of a very simple *separation* system in which a fire under the hydrocarbon mixture provides enough heat energy to vaporize part of the liquid in the first tank. As the mixture leaves the first tank, cooling water is injected, causing vapor to condense in the second tank. The condensed liquid in the second tank is richer in the light component, while liquid in the first tank is richer in the heavy component.

The open flame is an obvious fire hazard. Although a fired heater might be used, either a heat exchanger or a *reboiler* is a better source of heat, as shown in figure 1.12. In a heat exchanger, the liquid mixture to be heated is in the shell and steam is in tubes. A reboiler might have the liquid mixture in the tubes or in the shell, and can use either steam or hot oil for heating.

A simple separating system is improved by the addition of a *shell-and-tube* exchanger, called a *condenser,* that cools the mixture as it leaves the first tank, as illustrated in figure 1.13.

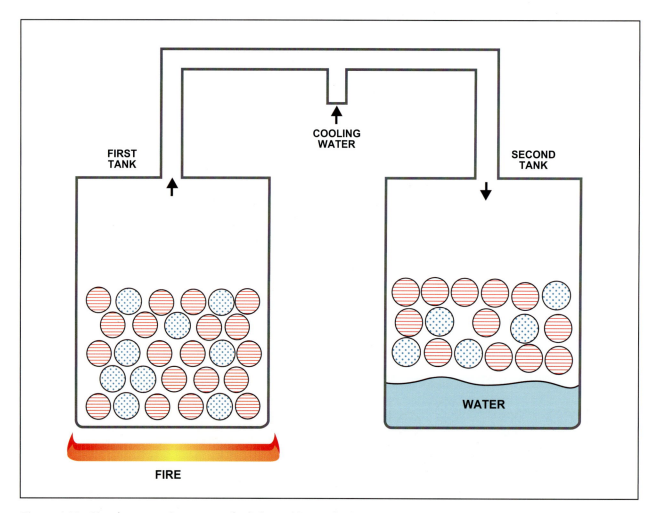

Figure 1.11   Simple separation system for light and heavy fluid components

Plant Processing of Natural Gas

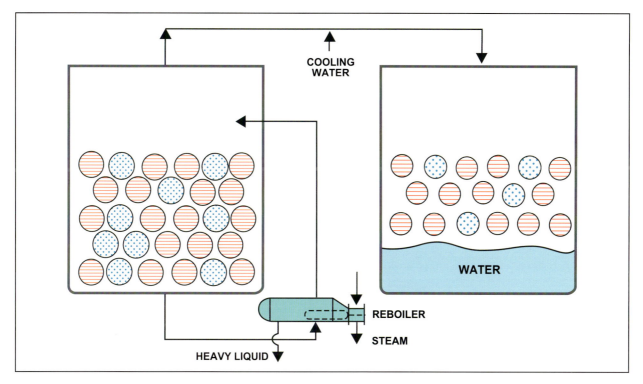

Figure 1.12   Simple separation system with reboiler

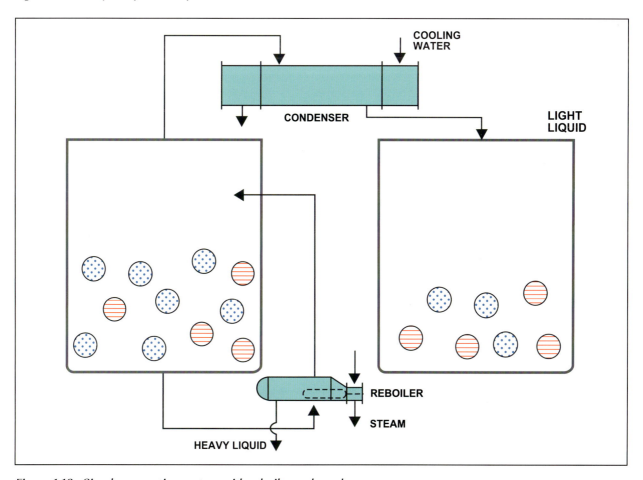

Figure 1.13   Simple separation system with reboiler and condenser

The condenser is more efficient than injecting cooling water directly into the process, which requires removing water from the condensed liquid later. The cooling water runs through the tubes and the mixture through the shell, condensing the vapor in the shell without coming into contact with it.

Even with those design additions, the basic separating system in figure 1.13 is still not very effective or efficient. It leaves light components in the heavy liquid and heavy components in the light liquid. Several tanks in series, as shown in figure 1.14, improve the separation efficiency. However, the changes and additions make the separating system complicated to operate and expensive to build.

A more efficient separation solution that reduces the amount of equipment is a *tower with trays*. There are several kinds of trays: *bubble cap,* valve, sieve, and perforated, among others. All provide good vapor and liquid contact, improving the heat exchange and separation efficiency. Trays used in towers are illustrated in figure 1.15.

Inside a *separation tower,* liquid flows across each tray and down a spout to the next lower tray. Meanwhile, vapor flows up the tower, bubbling through each tray of liquid. The trays are usually spaced about two feet apart, and towers can be designed to contain many trays (fig. 1.16).

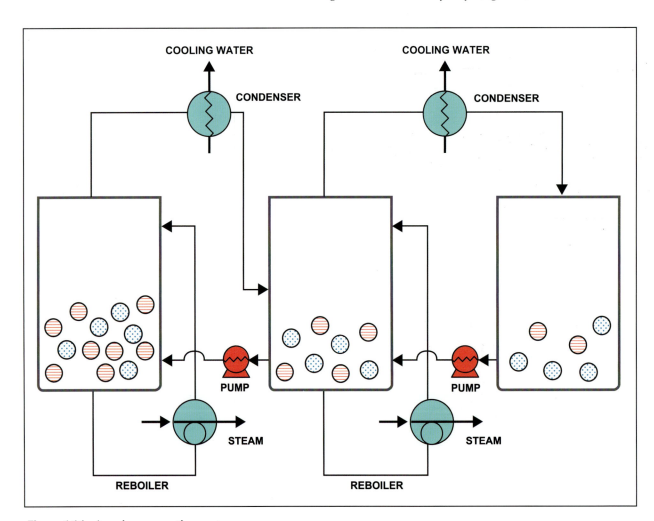

Figure 1.14 A series separation system

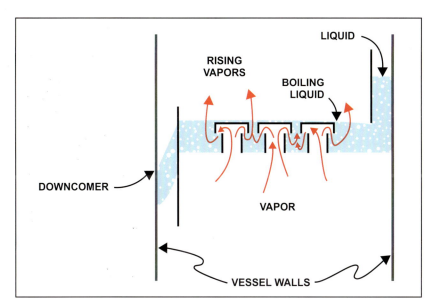

Figure 1.15   Bubble-cap tray for separation tower

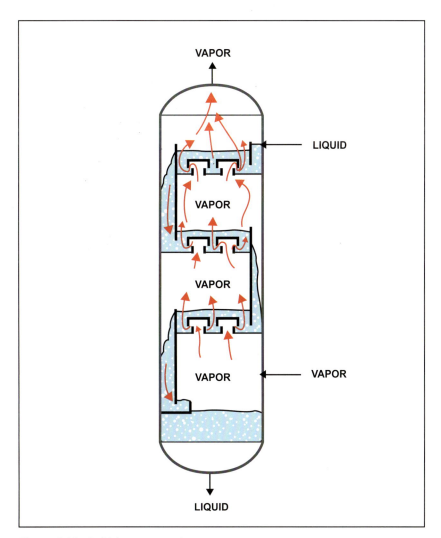

Figure 1.16   Bubble-cap trays in tower

Figure 1.17 shows a separation system using a tower with trays. The reboiler vaporizes the liquid at a temperature high enough so that a small amount of the light component is left in the heavy liquid. Except for the *heavy-liquid product,* all of the liquid that gets to the reboiler is vaporized. This vapor starts up the tower and contacts the liquid coming down the tower. The hot vapor heats the cold liquid, and the cold liquid cools the hot vapor. Hot vapor also vaporizes or *strips* some of the light components from the liquid. For this reason, the bottom section of the tower between the feed and reboiler is called the *stripping section.*

However, vapor flowing up the tower and past the point where the feed enters still contains heavy components that must be kept out of the light-liquid product. To keep out heavy components, part of the light-liquid product is pumped to the top tray of the tower. This stream is called *reflux.*

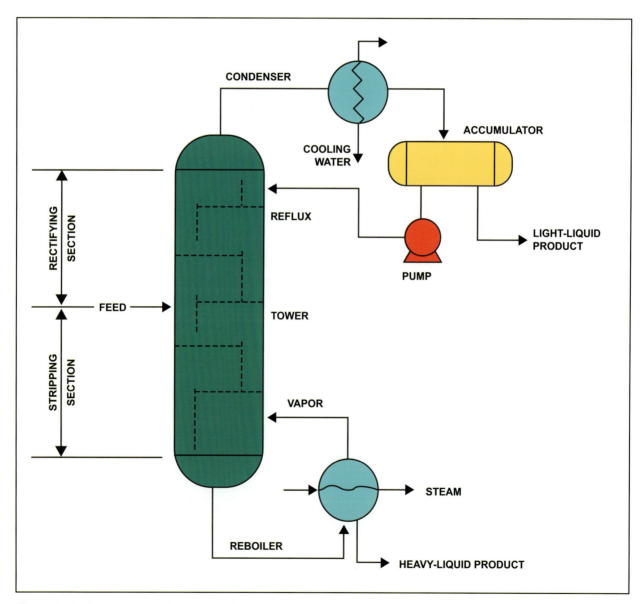

Figure 1.17   A tower separation system diagram

Reflux purifies the light-liquid product by cooling the vapors in the tower's top section and condenses the heavy components. The tower's top section above the feed is called the *rectifying,* or *rectification section.* These are the basic elements of a good separation system, and every gas processing plant uses these concepts.

REFERENCES

GPSA Engineering Data Book, 12th Edition, Gas Processors Suppliers Association, Tulsa, Oklahoma (2004).

GPA publication 2145[27]

TREATING AND PROCESSING

Plant unit configurations vary depending on the type of components of the feed gas and the final products desired for consumer use (fig. 2.1).

Courtesy of Chevron

Figure 2.1   Gas processing plant

Feed gas from various gas fields enters the gas plant through pipelines and goes through several units of *treating* and processing, as shown in figure 2.2. The main units perform the following functions:

- Remove oil and condensates
- Remove water
- Separate the natural gas liquids from the natural gas
- Remove sulfur and carbon dioxide
- Remove impurities such as mercury, oxygen, and BETX (benzene, ethylbenzene, toluene, and xylenes)

The first treating unit is the feed gas-receiving system and the condensate stabilization system. Condensate is a light hydrocarbon liquid obtained by condensation of hydrocarbon vapors. It consists of varying proportions of propane butanes, pentanes, and heavier components with little or no methane or ethane. The feed gas receiving system separates the feed gas into gases, aqueous liquid, and hydrocarbon liquid for further processing at plant units *downstream* (fig. 2.3).

The condensate stabilization system removes the light components such as methane, ethane, and propane, dissolved in the hydrocarbon liquid from the feed gas reception system (fig. 2.4). Hydrocarbon liquid normally contains a large amount of dissolved light components because of high pipeline pressures. These light components need to be removed to meet condensate product and other downstream processing requirements.

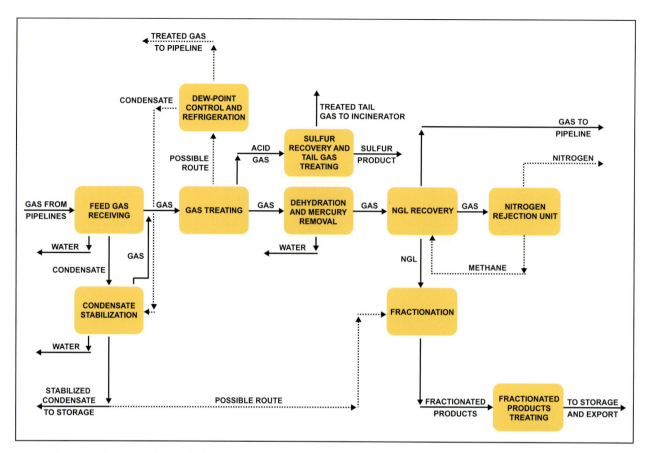

Figure 2.2 The functional units inside a gas processing plant

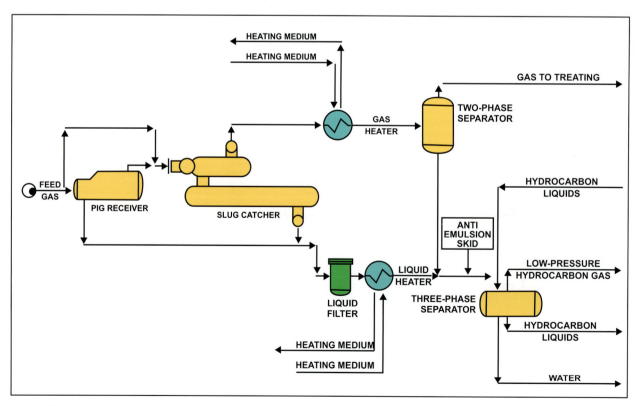

Figure 2.3 A typical feed gas receiving system

The feed gas can contain hydrocarbon liquids, aqueous liquids, hydrocarbon and acidic gases, entrained solids, and chemicals. Chemicals are often injected into the raw feed gas to control *corrosion* or to prevent formation of hydrates or solids. The feed gas can be dry or wet depending on the degree of *dew-point control* used *upstream* of the pipelines. Unless the pipelines are designed to be dry with no water and hydrocarbon liquids, liquids will accumulate in the pipelines and must be removed periodically.

The major feed gas reception system equipment includes *pig receivers*, *slug catchers*, feed gas heaters, and separators. The gases separated by the slug catcher are heated to prevent hydrate formation and to ensure they will remain as single gas phase after pressure letdown across control valves downstream. Pressure letdowns can create *retrograde condensation* problems of two-phase flow and liquid slugging. The gases are metered shortly after heating. Control valves in a series are used for pressure letdown and flow control. A *flow-control valve* can also be used for pressure control during start-up of the gas plant. After the pressure letdown, the gases flow to another separator to prevent any liquid carryover into the downstream gas treatment unit.

The liquids separated by the slug catcher are filtered, heated, and undergo a pressure letdown before entering a three-phase separator downstream.

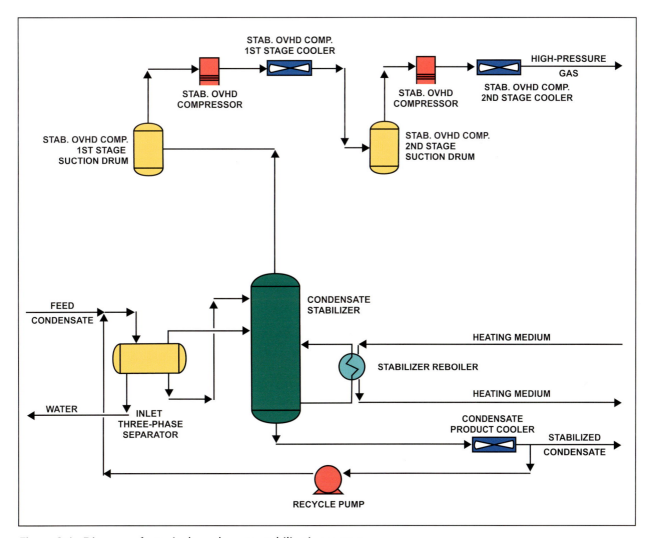

Figure 2.4   Diagram of a typical condensate stabilization system

The aqueous portion of the separated liquid is sent to waste water treatment or to chemical regeneration. Regeneration is a process that allows the chemical to be reused. The hydrocarbon portion of the separated liquid is sent to the condensate stabilization system. The hydrocarbon vapors released from heating and pressure letdown of the hydrocarbon liquid rejoin the processed gases separated by the slug catcher. The combined gases are then sent to the *gas treatment unit* along with the stabilizer overhead hydrocarbon gas returned from the condensate stabilization system.

In the feed-receiving system, the slug catcher and the three-phase separator send hydrocarbon liquid to another three-phase separator in the condensate stabilization system. Then, the collected liquid hydrocarbon goes to the stabilizer. The stabilizer is typically a reboiled *fractionation column* that gives a bottom stabilized condensate product and an overhead hydrocarbon gas stream.

The overhead hydrocarbon gas is compressed in multiple stages to the pressure of the gases separated by the slug catcher. The pressurized overhead gas is then cooled by a discharge cooler and rejoins the processed gases in the gas-treating unit. The overhead hydrocarbon gas can be used as a low-pressure fuel gas before compression and as high-pressure fuel gas after compression. When the compressor is down, the overhead gas is sent to the low-pressure fuel gas system.

The stabilized condensate product from the stabilizer is cooled and possibly combined with condensate products from the fractionation unit and sent to the condensate storage tank before being pumped out for sale.

DESIGN BASIS AND SPECIFICATIONS FOR TREATMENT UNITS

Feed Gas Basis

The field feed gas comes to the gas processing plant through pipelines from individual wells using different gas-gathering systems. The hydrocarbon content and richness of this feed gas will vary. It is critical that the engineers study the components and behavior of the gas in the various processing operations based on three considerations: the average feed, the rich feed, and the lean feed. Also, the water content, the feed conditions, temperature, and pressure are important factors.

In preparing plant *design cases*, engineers must work with pipeline and gas-production suppliers to effectively handle the feed gas flow, product composition, and removal of impurities. Design cases are based on specific conditions and the composition of the gas feed. Examples of gas feed design cases are shown in Table 2.1.

In plant design, the gas-receiving system is usually based on the rich/winter feed gas conditions. Leaner feed is the basis that defines the *turndown ratio* requirement of this unit. Turndown ratio is calculated by dividing the maximum system output by the minimum output at which controlled, steady output can be sustained. Cyclo-hexane and *aromatics*, such as benzene, toluene, xylene, and ethyl-benzene, are identified because of their tendency to freeze and form solids at very cold temperatures downstream. Also considered in the design case is the amount of sulfur-containing compounds, such as *mercaptans* and sulfides, and the specific treatment required to handle these compounds.

Table 2.1
Typical Gas Feeds

Gas Composition	Units	Average Feed	Rich Feed	Lean Feed
Carbon Dioxide	Mole %	1.490	1.090	1.450
Hydrogen Sulfide	Mole %	0.002	0.030	0.005
Nitrogen	Mole %	0.610	0.270	0.740
Methane	Mole %	83.000	77.900	85.800
Ethane	Mole %	7.100	8.000	11.400
Propane	Mole %	4.290	5.580	0.380
i-Butane	Mole %	0.690	0.880	0.010
n-Butane	Mole %	1.500	1.900	0.020
i-Pentane	Mole %	0.400	0.600	not present
n-Pentane	Mole %	0.300	0.500	not present
n-Hexane	Mole %	0.310	1.740	not present
n-Heptane	Mole %	0.080	0.460	not present
n-Octane	Mole %	0.040	0.240	not present
n-Nonane	Mole %	0.020	0.090	not present
Cyclo-Hexane	Mole %	0.036	0.036	0.036
Benzene	Mole %	0.018	0.018	0.018
Toluene	Mole %	0.019	0.019	0.019
Xylenes	Mole %	0.007	0.007	0.007
Et-Benzene	Mole %	0.001	0.001	0.001
Methyl Mercaptan	Mole %	not present	0.0002	not present
Ethyl Mercaptan	Mole %	not present	0.0001	not present
Di-methyl Sulfide	Mole %	not present	0.00005	not present
Isopropyl Mercaptan	Mole %	not present	0.0004	not present
Water	Mole %	0.007	0.432	0.007
Total	**Mole %**	**100**	**100**	**100**

Product Specifications

Hydrocarbon Treating Unit

From the feed gas reception and condensate stabilization unit, the processed gases flow to the *hydrocarbon treating* unit. The minimum temperature limit of the gas treatment unit is determined by the requirements of the technology selected and specified by licensors or vendors.

Condensate Product Storage Tanks

The condensate product in the storage tanks must meet a *Reid Vapor Pressure (RVP)* requirement and butane content specifications. Typical specifications are given as follows:

- Reid Vapor Pressure: 10.0 psia at 100°F per ASTM D-323

- Butane content: 0.5 liquid volume % maximum

- If a fractionation plant is downstream, the condensate product might be delivered to either a *debutanizer* or a *depentanizer.*

EQUIPMENT SELECTION AND DESIGN

In designing the pig receiver and the slug catcher, the pressure capacity must be the same as that of the pipeline.

If *high integrity pressure protection system (HIPPS)* isolation valves are used, lower downstream design pressures may be used. However, plant designs using HIPPS require a thorough knowledge of applicable government regulations and standards, industry codes, and follow recommended system practices (Summer, A.E., 2000).

Suggested alternatives to pressure relief devices, such as HIPPS, are provided by the *American Petroleum Institute (API)* 521 and Code Case 2211 in the *American Society of Mechanical Engineers (ASME)*, Section VIII, Divisions 1 and 2. Both of these organizations are leaders in developing standards for the gas industry. These instrumented systems are called *safety instrumented systems (SIS)* because their failure can result in the release of hazardous chemicals and/or cause unsafe working conditions. Using such systems requires meeting or exceeding the protection provided by the pressure-relief device such as HIPPS.

To have a correctly sized separator or slug catcher, the size of incoming liquid slugs must be determined. The best references for calculating slugs generated by inlet *flow rate* or pigging frequency are given in Section 17, pages 17–21 of the *GPSA Engineering Data Book*, 12th edition, 2004.

Pig Receivers

A *pig receiver system* is required for a pipeline unless the pipeline transfers only dry gas and hydrocarbon liquids (figs. 2.5 and 2.6).

The function of a pig receiver system is to periodically remove the accumulation of liquids at low spots in a pipeline. A periodic removal of these liquids prevents corrosion, improves pipeline efficiency by reducing pressure losses, and limits maximum slug size. Each pipeline might have its own pig receiver or a combination launcher/receiver (fig. 2.7). The size of the liquid accumulation or slug is determined by pipeline temperature, topography, pipeline length, and raw feed gas composition.

Courtesy of Girard Industries

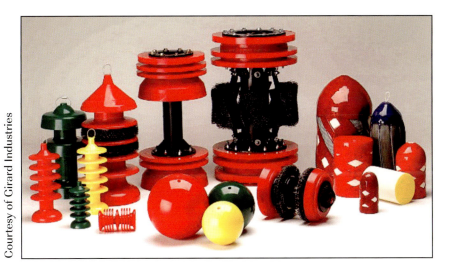

Figure 2.5 Various types of pigs used for pipeline cleaning

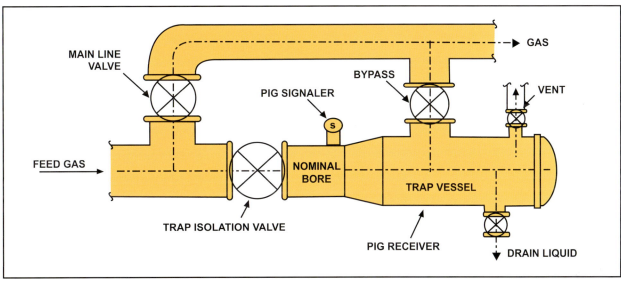

MAIN LINE
VALVE
GAS
BYPASS
PIG SIGNALER
S
VENT
FEED GAS
NOMINAL
BORE
TRAP VESSEL
TRAP ISOLATION VALVE
PIG RECEIVER
DRAIN LIQUID

Figure 2.6 A typical pig receiver system

Courtesy of Piping Technology & Products, Inc.

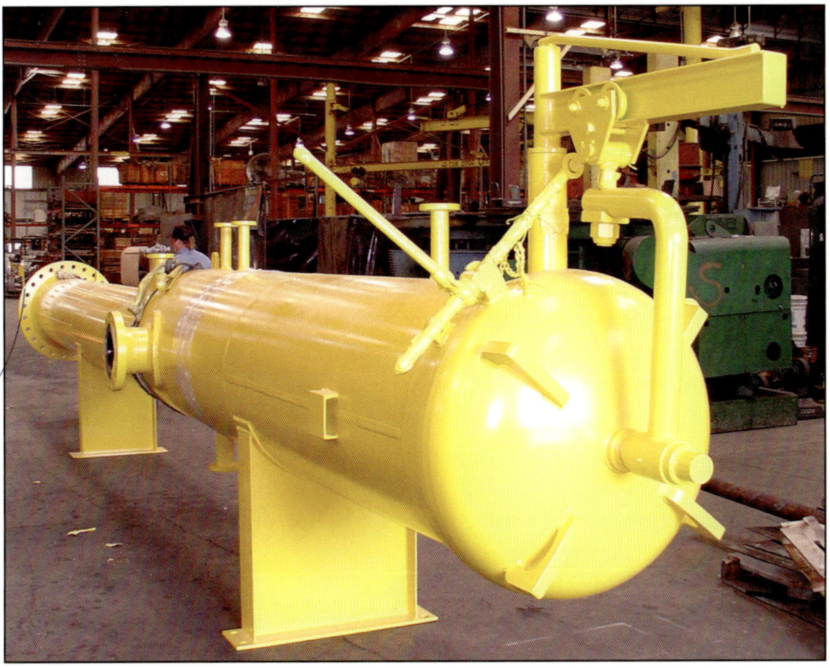

Figure 2.7 Pig launcher/receiver

Slug Catchers

The function of slug catchers is to separate liquid slugs formed in pipelines. Slugs form when:

- liquid waves grow large enough to bridge the diameter of the pipe;
- liquid accumulates at the low spots of the pipe;
- increase feed flow rates result in excess liquids;
- *pigging* pushes out the entire accumulated inventory of liquids in the low spots of the pipeline.

A slug catcher might be either a vessel or finger-type device constructed of pipes (fig. 2.8). For high-pressure servicing, pipes are used rather than the less cost-effective thick-wall vessels. Pipes allow for easier addition to increase the slug catcher capacity.

If time is needed for gas-liquid separation, the slug catcher size should be increased accordingly. An average-sized separator is required during initial start-up and any restart after pipeline shutdown if the pipeline is designed to handle dry gas only.

Courtesy of StatoilHydro. Photo by Torstein Tyldum.

Figure 2.8   Finger-type slug catcher

The maximum size of the slug produced by pigging must be considered in equipment design. A slug catcher is designed to hold several weeks' accumulation resulting from the weekly piggings. If more frequent pigging is scheduled, the size and accumulation of the slugs is reduced (fig. 2.9).

Other factors affecting the slug catcher design are gas temperature, pipe length, topography, and gas composition. Cold temperatures, longer pipes, changes in topography, and richer gas all result in more liquid slugs. Liquid slug compositions can vary from 100 vol% hydrocarbons to 80 vol% water. During increased production, liquid slugs consist primarily of water and chemicals such as hydrates and corrosion inhibitors. Design of slug catchers varies according to the requirements of the processing plant (figs. 2.10, 2.11, and 2.12).

Courtesy of StatoilHydro. Photos by Torstein Tyldum.

Figure 2.9   Slug catcher being transported

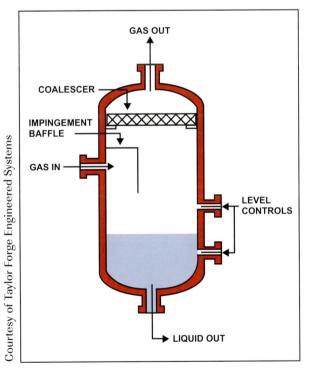

Courtesy of Taylor Forge Engineered Systems

Figure 2.10   Vertical slug catcher vessel

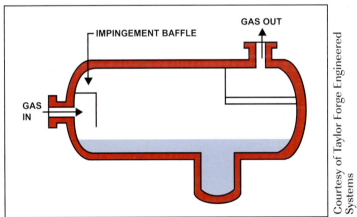

Courtesy of Taylor Forge Engineered Systems

Figure 2.11   Horizontal slug catcher vessel

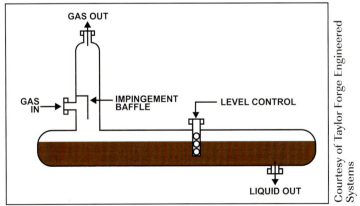

Courtesy of Taylor Forge Engineered Systems

Figure 2.12   Pipe-fitting-type slug catcher

Feed Gas Receiving and Condensate Stabilization

Condensate Stabilizers

Hydrocarbon liquids can be stabilized by *flashing*, which quickly vaporizes liquids to lower pressures in a series of heaters and separators. However, a *stabilization column* with a reboiler and liquid feed from the top produces a higher quality and better-controlled final product. Large volumes of hydrocarbon liquids can be processed in a stabilization column.

When stabilization columns are trayed instead of packed, column or towers are usually selected for maximum turndown flexibility. Bubble-cap towers can be used for the highest turndown ratio of 10:1 or 10%. Valve-tray towers can be used for 30%–40% turndowns. Packed towers can be used when turndowns are not a concern.

The stabilizer column temperature at the bottom should be less than 400°F to avoid degradation of heavy hydrocarbon components. High bottom temperature might also create a situation where water cannot leave from the bottom and, thus, accumulates in the column. Placing a separator before the tower or having a middle water draw tray prevents water from recirculating and accumulating in the column. Some condensate stabilizers are designed for 150% or more of equilibrium liquid to take care of slugs.

Condensate Stabilizer Reboilers

The condensate stabilizer reboiler can be designed for two situations. When pipeline topology indicates large amounts of hydrocarbon liquid accumulations, two 70% design capacity reboilers are recommended to ensure a high on-stream factor for overall gas processing. When the amount of hydrocarbon liquids to be processed by the condensate stabilizer is small, a single 100% design capacity reboiler, with no spare, is recommended because the condensate stabilizer can stop taking new feed and allow hydrocarbon liquids to accumulate in the slug catcher until the single reboiler comes back online.

Stabilizer Overhead Compressors

Reciprocating compressors have high maintenance requirements. However, they are recommended for low-flow rates, high pressure, when overhead gas composition or molecular weight changes frequently, and when centrifugal compressors have problems staying on-line. If a *centrifugal compressor* is used and the compression ratio is high, an *interstage cooler* should be used to protect the seals from a high discharge temperature over 350°F (177°C). This upper temperature limit must meet the compressor manufacturer's specifications.

One drawback of an interstage cooler is that the circulation of propane requires a larger condensate stabilizer and a larger compressor. Propane liquid from the interstage cooler is separated from the stabilizer overhead gas in a knock-out drum and returned to the condensate stabilizer. Normally, a discharge cooler is needed for start-up to remove the heat generated by recirculation.

Gas and Liquid Heaters

The feed gas reception unit uses heaters to raise the temperatures of the feed gas and liquids separated by the slug catcher. The design of the feed gas heaters should:

- prevent gas hydrate formation;
- prevent retrograde condensate formation;

- ensure that the feed gas temperature does not go below the minimum design temperature of equipment downstream;
- raise the feed gas temperature to the required temperatures of the gas-treating unit.

Gas with higher CO_2 concentrations will cause a kinetic reaction and generate more heat. In the design of the feed gas heater, the required temperature at low CO_2 concentrations must be considered.

A heater is used to heat the hydrocarbon liquid to prevent freezing after depressurization before going into the condensate stabilizer. The amount of heating should be optimized to reduce the demand on the stabilizer reboiler.

The problems solved by the gas feed reception and condensate stabilization unit equipment are:

- The pig system is used to periodically remove liquids accumulating in the feed pipelines.
- The slug catcher is used to separate the liquid slugs in the feed gases.
- The condensate stabilizer removes the light components so that the condensate product can meet specifications.
- The gas and liquid heaters prevent gas hydrate formation and liquid freezing.

REFERENCES

GPSA Engineering Data Book, 12th edition, Gas Processors Suppliers Association, Tulsa, Oklahoma (2004).

Summer, A.E., "Instrumented Systems for Overpressure Protection," AIChE Chemical Engineering Progress issue of November (2000), Wiley Press, New York, New York.

PROCESS DESCRIPTIONS

Raw gas comes from production fields through the pipelines to the feed gas receiving unit and condensate stabilization unit. The raw gas then flows to the gas-treating unit and then to a dew-point control and refrigeration unit or a *natural gas liquid (NGL)* recovery unit. An export compression system is sometimes used after the dew-point control unit to pressurize the gas to the requirements of the *pipeline grid*. Finally, the gas is sent to the consumers through a pipeline grid.

A dew-point control unit helps to prevent liquid condensation in the pipeline grid under various pressures and temperature conditions. There are two kinds of dew-point control required: a *water dew-point control* and a *hydrocarbon dew-point control*. In water dew-point control, there are several dehydration, or water removal, methods available, including the *silica gel*, glycol, and *molecular sieve*. Hydrocarbon dew-point control also has various methods available including refrigerated *low-temperature separation (LTS)*, *expander, Joule-Thomson (J-T)*, and silica gel. Companies might use glycol gas dehydration for water dew-point control and a refrigeration cooling system for hydrocarbon dew-point control. More explanation of gas dehydration is given in Chapter 6.

The purpose of a *refrigeration system* is to remove heat from the feed stream in a heat exchanger. Heat exchangers are referred to as *evaporators* or *chillers* and provide the required cooling level for various gas processing applications. Refrigeration systems use refrigerant, called *working fluid*. Working fluid is selected based upon temperature requirements, availability, economics, and previous experience. The availability of ethane and propane on hand at natural gas processing plants makes these gases the prime choice as working fluids. In gas plants, propane is normally the preferred refrigerant.

COST ESTIMATE

Dew-point depression is defined as the difference in degrees between the feed gas temperature and the dew point of a fluid. The dew-point depression difference in degrees determines the best process to use for dew-point control.

Depending on the amount of dew-point depression required, an economic evaluation is done to compare installation costs and operating costs for the various processes. Using an average 80°F (45°C) dew-point depression requirement, there are several processes available for the dew-point control. Three of the most widely used process options are the silica gel process, the glycol/propane refrigeration process, and the glycol/J-T valve cooling process.

3
Dew-Point Control and Refrigeration Systems

Silica Gel Process

The *silica gel process* is used for water as well as hydrocarbon dew-point control. In this process, water and heavy hydrocarbons are removed from the feed gas on multiple *bed*s of silica gel adsorbent (fig. 3.1).

As shown in the flow diagram, feed gas enters the feed separator to knock out any liquid. The gas is then divided into two streams. The larger stream (more than 50% of the stream) is depressurized, mixed with the cooled and leaner *regeneration gas*, and routed to the two absorbers in an *adsorption* mode. The gas flows from the top of the bed and exits the bottom.

After passing through the adsorber after-filter, the product is then routed to the export compressor suction *KO drum* (knockout drum). The KO drum is a separator used to separate the gas and liquid. The gas is then compressed by the export compressor and cooled by the export compressor discharge cooler for export through the pipeline. All three parts in the export compression *train* are provided with a 100% spare or standby unit. Plant operations will put the standby unit online whenever there is a halt in the operating unit.

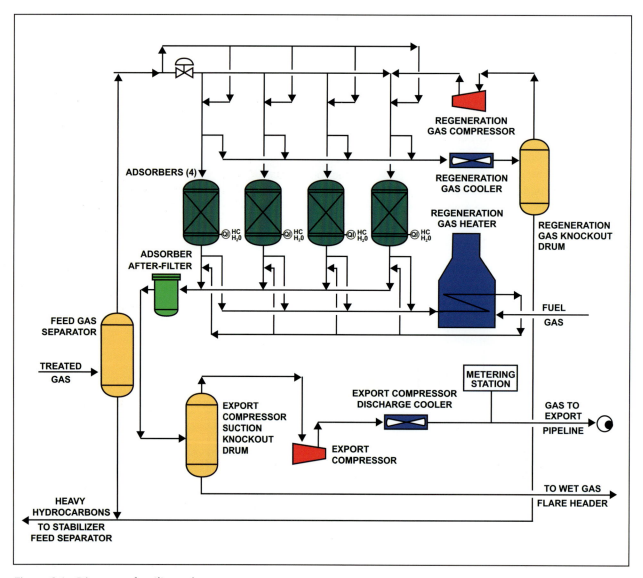

Figure 3.1   Diagram of a silica gel process

The smaller stream (less than 40% of the gas stream) is used to cool the freshly regenerated or restored bed. Then, the gas is heated in the regeneration gas heater. Heated gas enters at the bottom of the bed, which is ready for regeneration. As the hot regeneration gas flows up the bed, moisture and heavy hydrocarbons are *desorbed* (removed from) and separated from the adsorbent. The gas is cooled by the regeneration gas cooler. Water and heavy hydrocarbons condensate are separated from the gas in the regeneration KO drum. The liquid stream from the regeneration KO drum is routed to the condensate stabilization unit. Vapors are compressed and mixed with the larger portion of the feed stream and routed to the two beds operating in adsorption mode.

A total of four beds is recommended by the equipment vendors based on their licensed processes. Two beds operate in adsorption mode, one in cooling, and the remaining one in regeneration mode. The product gas quality is determined by the length of time the gas is on the adsorbent.

Glycol/Propane System

In the *glycol/propane system,* glycol is used for water dew-point control, and the refrigeration system is used for hydrocarbon dew-point control. Several different kinds of glycols can be used such as *methyl ethylene glycol (MEG), ethylene glycol (EG), diethylene glycol (DEG), or triethylene glycol (TEG).* The type of glycol used depends on the processing plant's determination of what is most economical and efficient for their product (fig. 3.2).

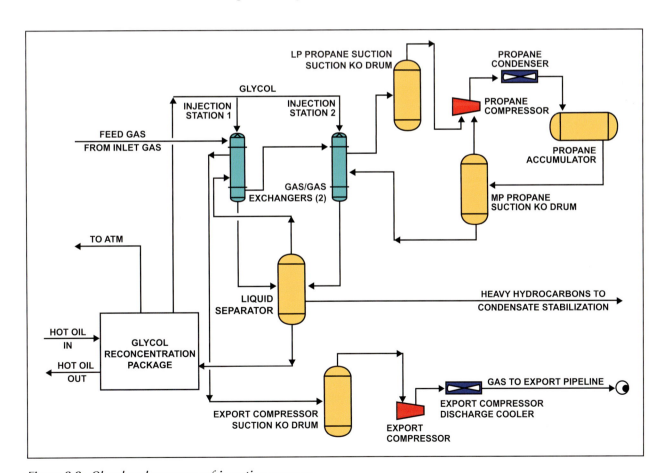

Figure 3.2   Glycol and propane refrigeration process

In the *glycol cryogenic process,* the feed gas water content is lowered by treating it with a circulating stream of glycol to remove moisture. The gas is further cooled, condensing the heavy hydrocarbons to achieve the desired hydrocarbon dew point.

Feed gas enters the vertical *gas/gas exchanger* on the tube channel side of the heat exchanger. The gas is sprayed with an approximately 80% glycol-and-water mixture to remove water from the feed gas. The gas-and-liquid mixture exiting the tubes is separated into glycol and liquid hydrocarbons in the bottom tube channel cover. The mixture is then drained into the *liquid separator* by a *level control* valve station.

The gas is cooled further in the *cooler,* a vertical heat exchanger, by evaporating propane on the shell side of the heat exchanger. Glycol is sprayed in the gas stream at the cooler tube sheet entrance. As gas flows down the tubes, heavy hydrocarbons condense. The residual moisture is picked up by the glycol stream. The cooled gas, condensed hydrocarbon, and glycol mixture exit the bottom of the cooler and are drained to a liquid separator. A vertical vessel called a liquid separator separates the diluted glycol and hydrocarbon streams. The glycol concentration, spray rate, and condensing temperature control the product gas, water, and hydrocarbon dew points. The gas exiting the liquid separator is heated in the gas/gas exchanger by cooling the incoming feed gas stream. Gas is then routed to the export compressor suction KO drum, compressed by the export compressor, and cooled by the export compressor discharge cooler for export via the pipeline. All three items in the export compression train are provided with a 100% spare or standby unit.

A liquid stream containing glycol and heavy hydrocarbons is also separated in the liquid separator. The heavy hydrocarbon stream is sent to the condensate stabilization unit. The glycol stream is regenerated and restored to its original concentration by heating it in a *reconcentrator* to evaporate moisture collected from the feed gas. Glycol is then recirculated to the two heat exchangers.

A *closed loop* propane refrigeration system is used to supply the coolant to the gas cooler.

Glycol/J-T Valve Cooling Process

Glycol is used to control the water dew point while the *Joule-Thomson (J-T) valve system* controls the hydrocarbon dew point. Different glycols can be used for this process such as MEG, EG, or TEG depending on the processing plant's economic evaluation and preference.

In this process, the feed gas water content is lowered by treating it with a circulating stream of glycol to remove moisture and by using J-T effect cooling to condense the heavy hydrocarbons (fig. 3.3). Feed gas enters the vertical gas/gas exchanger in which the feed gas enters the tube channel. There, it is sprayed with an approximately 80% glycol-and-water mixture for water removal from the feed gas. The gas-and-liquid mixture exiting the tubes is disengaged and separated in the bottom tube channel cover of the heat exchanger. The glycol/hydrocarbons stream is then drained into the liquid separator by a level control valve.

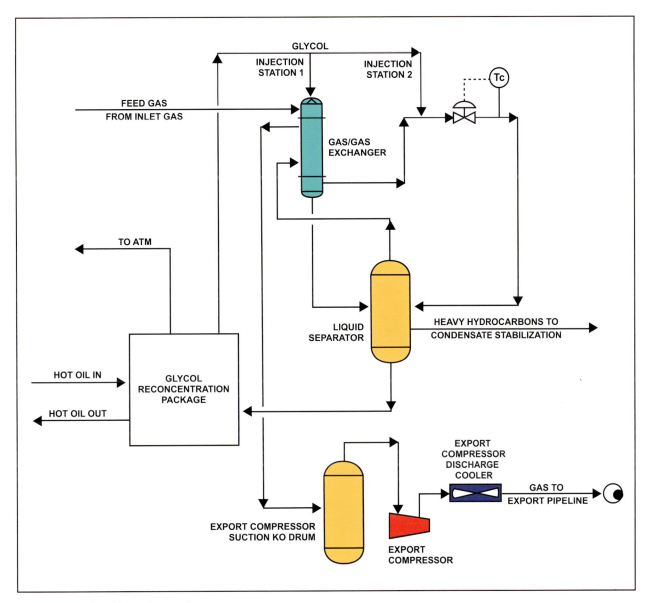

Figure 3.3   Glycol/J-T valve cooling process

The process gas stream then flows through a valve to provide J-T effect cooling. As a result of lowering the feed gas temperature, the heavy hydrocarbons condense. A small amount of glycol spray at the inlet of the pressure letdown valve prevents any hydrate formation. The cooled gas stream containing condensed hydrocarbons are drained into the liquid separator. The liquid separator is a vertical vessel that separates product gas, rich glycol, and condensed hydrocarbon streams. The selected glycol concentration and spray rate, and the gas condensing temperature will control the product gas, water, and hydrocarbon dew points.

Gas exiting the liquid separator is heated in the gas/gas exchanger where it cools the incoming feed gas stream. Gas is then routed to the export compressor suction KO drum, compressed by the export compressor, and cooled by the export compressor discharge cooler for export via the pipeline. The export compression train is provided with 100% spare to achieve higher on-stream availability.

The liquid stream of glycol and heavy hydrocarbons is separated in the liquid separator. The heavy hydrocarbon stream is sent to the condensate stabilization unit. The glycol stream is restored to its original concentration by heating it in a reconcentrator to evaporate moisture collected from the feed gas. The glycol stream is then recirculated to the gas/gas exchanger and the J-T valve.

Comparison of Dew-Point Processes

The silica gel process option does not need any refrigeration system. In comparison to the glycol/propane refrigeration system or the glycol/J-T system, the silica gel process is usually the most expensive of the three options.

A J-T valve cooling process has lower capital cost, is easier to operate, and has fewer equipment pieces. The glycol/J-T valve cooling option does not require propane refrigeration.

However, the energy consumption for the glycol/J-T valve cooling option tends to be higher than the energy consumption for the glycol/propane refrigeration option. The operating costs are higher due to the higher fuel gas consumption needed for the extra power and for the control of increased emissions.

To achieve dew-point depression of about 27.0°F (15.0°C), advanced separator technology might be used. Shell's *Twister™ Supersonic separator* technology can treat produced gas at supersonic velocities, extracting water and hydrocarbon liquids simultaneously. Compared to conventional technologies, this process requires no chemicals, has no moving parts, and allows cost reduction, especially on offshore installations.

Unit Specifications

If gas from the dew-point control unit is needed for the national gas grid, the grid's specific requirements of water and gas hydrocarbon dew points must be met. Typically, the gas specifications might be:

- Water <48 mg/Sm³ (2.88 pound/MMscf)
- Gas hydrocarbon dew-point temperature <32°F (<0°C) at (2.5 to 8.7 MPa) 363 to 1263 psia

Unit specifications might also vary by country.

THE REFRIGERATION SYSTEM

A refrigeration system operates through the continuous circulation of a refrigerant in a four-step cycle: expansion, *evaporation,* compression, and condensation. Although a plant refrigeration system operates much like a home air conditioner, it uses a different refrigerant. Liquid propane flows from the *gas surge tank* to an oil chiller and the *gas chiller,* as shown in figure 3.4.

The gas and oil chillers perform the same function as an evaporator in a home air conditioner. In the home system, hot air blows across the evaporator, causing the refrigerant to vaporize. In a plant refrigeration system, the oil and gas are warmer than the propane refrigerant. As the oil and gas flow through the tubes of the gas chillers, the propane surrounding the tubes in the chiller shells vaporizes.

Propane vapor from the chillers flows through the suction scrubber and on to the compressor. The compressor raises the pressure of the propane vapor from the chillers, so the propane will condense when it is cooled, just as refrigerant is compressed and condensed in a home air-conditioning system. After the propane vapor is compressed, it can be condensed with cooling water or air coolers and returned to the surge tank as a liquid.

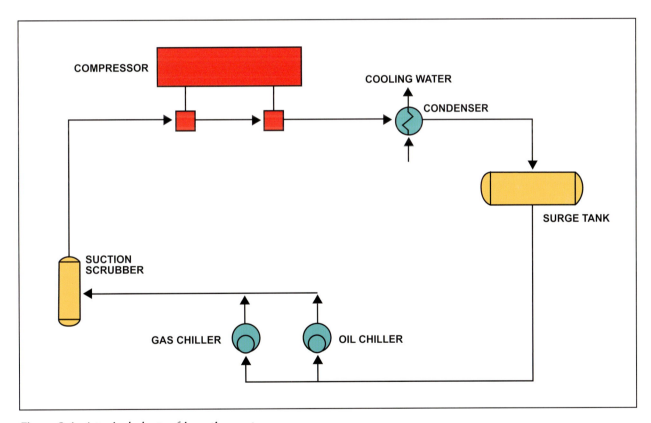

Figure 3.4 A typical plant refrigeration system

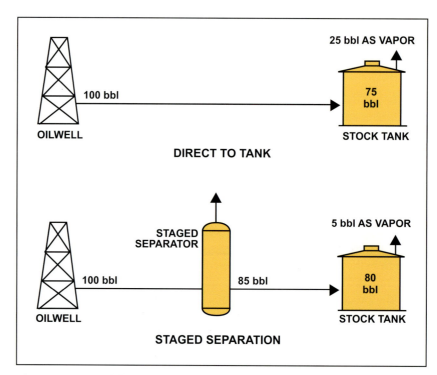

Figure 3.5   Two different layouts of simple-staged separation process

Economizers

A propane refrigeration system uses the energy of latent heat to change the propane from a liquid to a vapor in the chillers. The propane in the surge tank is similar to the liquids from a high-pressure oilwell. When oil is produced from the well and flows into a stock tank, some of the vapor from the oil leaves the top of the tank, where it is collected in a vapor recovery system. The liquid left behind is the produced oil.

One way to increase oil recovery in the stock tank is to produce the well through several separators in a series, a process called *staged separation* (fig. 3.5).

A process similar to staged separation can be useful in a propane refrigeration system of a gas processing plant (fig. 3.6).

In gas processing, the staged separator is known as an *economizer* because it saves propane circulation rates when propane is used as a refrigerant. Economizers let more of the surge tank propane reach the chillers as a liquid and they also conserve compressor horsepower.

Large, gas turbine-driven refrigeration compressors work by compressing a gas in stages (fig. 3.7). As the gas is compressed in each stage, it picks up heat. However, that heat can be dissipated by sending some of the propane liquid to the compressor between stages (or interstage) to serve as a cooling medium. Refrigeration systems might have several economizers, each at a different pressure, corresponding to the compressor's interstage pressures.

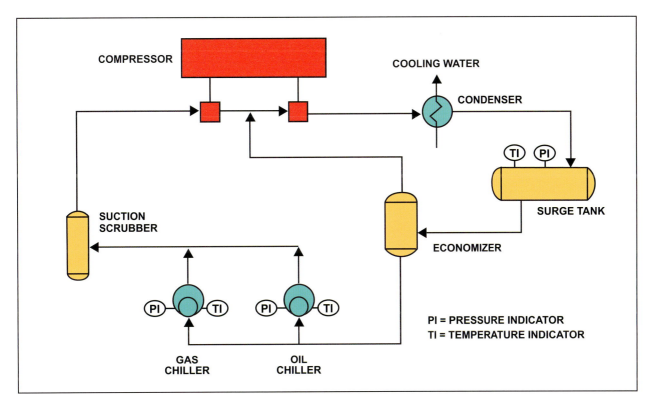

Figure 3.6   Diagram of a refrigeration system using staged separation

Figure 3.7   Gas turbine-driven propane refrigeration compressor in a natural gas plant

Dew-Point Control and Refrigeration Systems

Chillers

The gas or *lean oil* to be cooled flows through all the chiller tubes. For maximum refrigeration, all the tubes must be covered with propane. Space left at the top of the chiller shell allows the vapor to separate from the liquid and flow to the compressor without carrying liquid with it. See the chiller diagram in figure 3.8.

Notice that the level *controller* is set to maintain the liquid level just above the top of the tubes. As the liquid vaporizes, the level controller opens the control valve to let more liquid propane flow into the chiller.

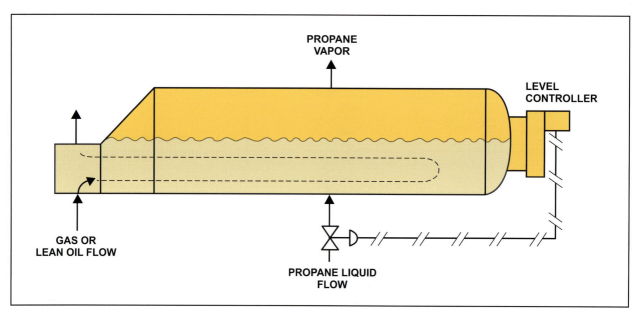

Figure 3.8 Chiller diagram

Possible Problems

Like most fluids, the propane used in plant refrigeration systems is not pure. It contains ethane and butanes. Small amounts are not a problem, but too much ethane in the propane will increase the compressor discharge pressure. Too much butane increases the temperature in the chillers.

As shown in figure 3.6, the refrigeration system employs staged separation. Note the *temperature indicators (TI)* and *pressure indicators (PI)* on the surge tank and the chillers. Operators use these gauges, along with a graph that compares condensing or chilling pressures and temperatures, to determine the quality of the propane in the system.

To determine how much ethane is in the propane, the pressure and temperature of the propane surge tank are recorded first. Then, the point corresponding to that pressure and temperature is located on a graph (fig. 3.9).

The point should be near one of the ethane percentage curves on the graph. Too much ethane increases the compressor's discharge pressure. If the ethane content of the propane is more than 2%, the excess should be vented from the surge tank.

To determine how much butane is in the propane, the pressure and temperature of the propane in the chiller are recorded. Then, that point is located on a graph (fig. 3.10).

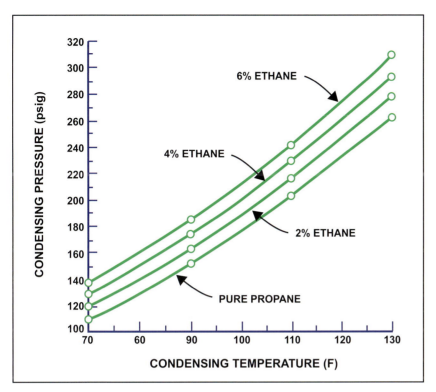

Figure 3.9   Graph used to determine amount of ethane in propane

At a fixed compressor suction pressure, butanes in the propane increase the chiller temperature. If the chiller temperature is too high, some of the butanes (a liquid at this point) should be drained from the chillers.

When the refrigerant condenser is operating efficiently, the compressor will run at a minimum discharge pressure. For air-cooled condensers, the fans, drivers, and *drive* assembly must be well-maintained to function efficiently.

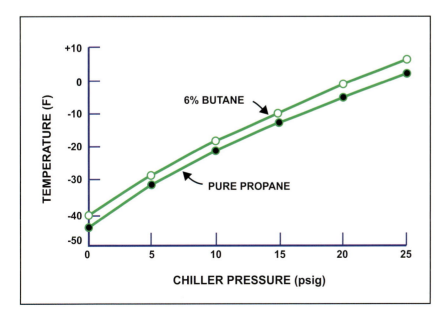

Figure 3.10   Graph used to determine amount of butane in propane

Cooling-water condensers allow for relatively simple monitoring of the condenser operation. The circulation to the condensers must be maintained and a record kept of two key temperatures: refrigerant from the condenser and the cooling water from the condenser. While recording the temperatures, the flow of refrigerant should be noted. If there is no refrigerant temperature gauge, some measure of the compressor load, such as the engine manifold pressure, should be recorded.

As the cooling-water scale builds up in the condensers, these two temperatures, at the same refrigerant flow or compressor load, will be farther apart. The condensers should then be cleaned before too much product recovery is lost due to an excessive compressor discharge pressure. When the temperatures are recorded, the refrigerant flow should be tracked and noted. If the refrigerant flow is not metered, some measure of compressor load, such as the engine manifold pressure, should be recorded.

Multiple-Stage Refrigeration

It is common to have two, three, and even four stages in a refrigeration system. In many lean-oil-based gas processing plants, multiple stages of refrigeration are required. Figures 3.11 and 3.12 are examples of two- and three-stage refrigeration systems.

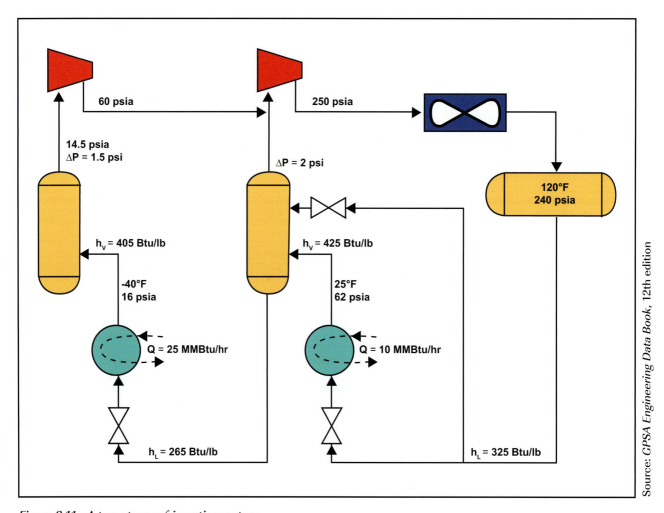

Source: *GPSA Engineering Data Book*, 12th edition

Figure 3.11   A two-stage refrigeration system

Plant Processing of Natural Gas

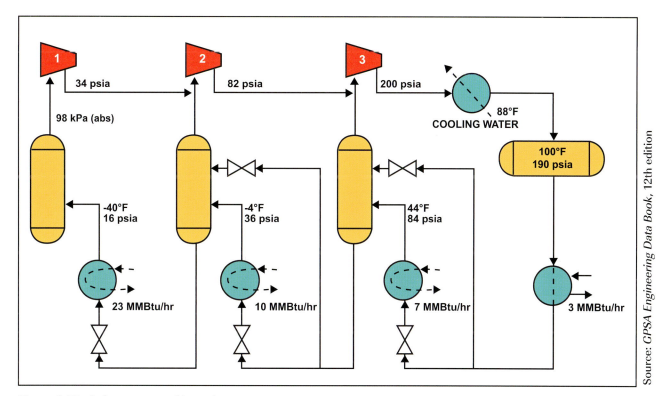

Figure 3.12   A three-stage refrigeration system

Figure 3.13 is an example of an ethane/propane cascade refrigeration system. In this system, compressed ethane is condensed with liquid propane as cooling water or fin-fan coolers cannot achieve the –25°F condensing temperature desired. As shown in figure 3.13, in an ethane/propane cascade refrigeration system, very low ethane evaporator temperature (–120°F in this case) can be achieved. Finally, figure 3.14 shows how the power requirements vary with evaporation and refrigerant-cooling temperatures.

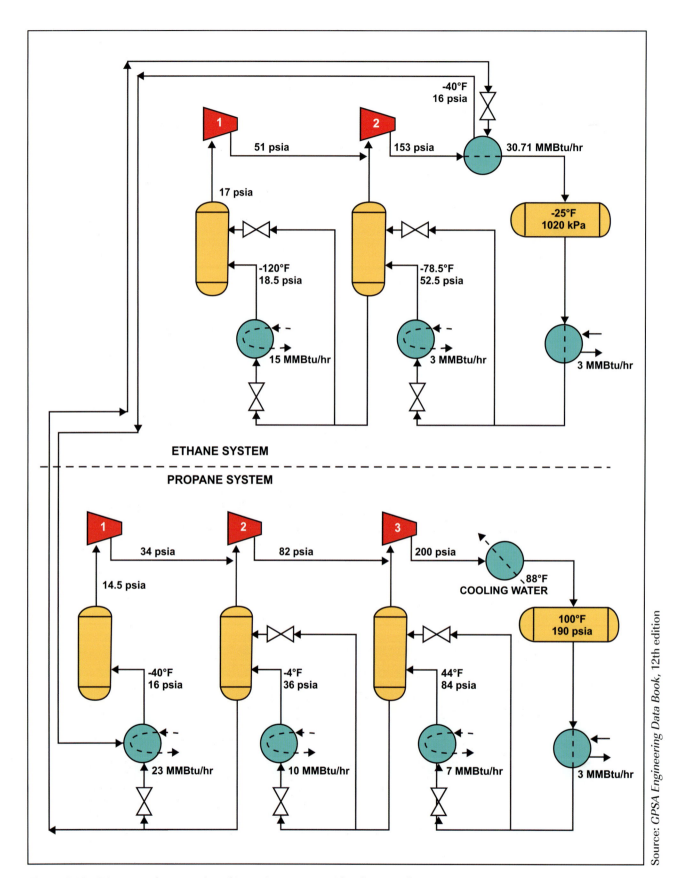

Figure 3.13   Diagram of a cascade refrigeration system with ethane and propane

Source: GPSA Engineering Data Book, 12th edition

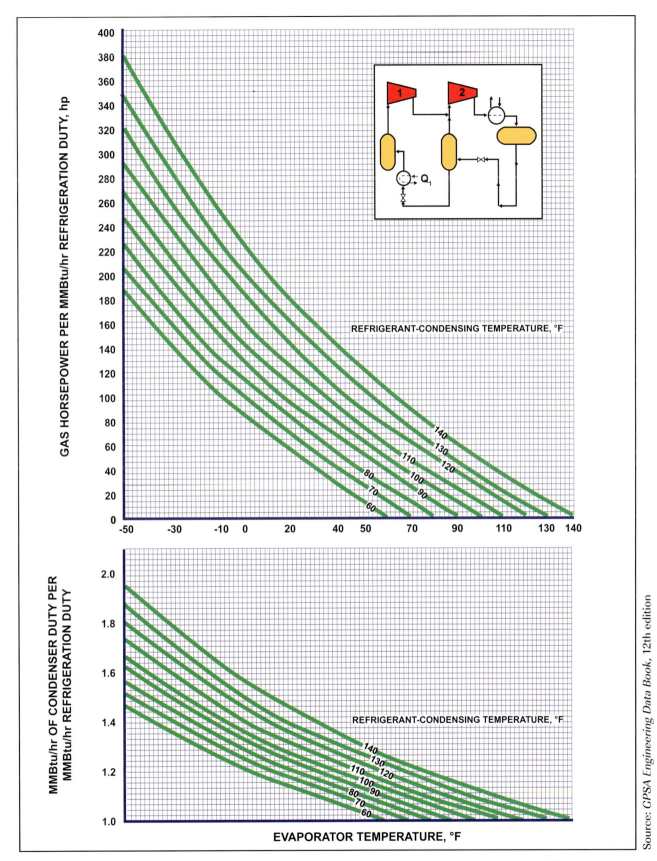

Figure 3.14   Two-stage propane refrigeration system (charts to calculate horsepower required at different evaporator and condenser temperatures)

REFERENCES

GPSA Engineering Data Book, 12th edition, Gas Processors Suppliers Association, Tulsa, Oklahoma (2004).

ASHRAE Handbook Fundamentals, American Society of Heating, Refrigeration and Air-Conditioning Engineers, Inc., Atlanta, Georgia (2001).

Epsom, H., Schinkelsshoek, P., Twister, B.V., "Shell-Engelhard Sordeco Process: Silica-Gel Process for Dew-Point Control, Supersonic Gas Conditioning for NGL Recovery," Offshore Technology Conference, Houston, Texas (May 1–4, 2006).

Summer, A.E., "Instrumented Systems for Overpressure Protection," AIChE Chemical Engineering Progress November (2000), Wiley Press, New York, New York.

GAS-TREATING PROCESSES

Hydrocarbon streams, both gaseous and liquid, might contain contaminants such as H_2S and CO_2 that must be removed before further processing and marketing. Removal of H_2S, CO_2, and other sulfur compounds, commonly called *acid gases*, is normally referred to as hydrocarbon treating or *sweetening*.

Treated gas regulations and specifications are stringent regarding residual H_2S and other sulfur species. Typical U.S. sales gas contracts restrict the following:

H_2S	< 0.25 grains/100 scf (about 4 ppmv)
Total sulfur compounds	< 5 grains/100 scf (about 80 ppmw)
CO_2	< 2 % mole

Acid gas components can be removed from a sour gas stream by:

- chemical reaction using liquids or solids;
- physical *absorption* in liquids;
- adsorption on solids;
- diffusion through membranes.

The acid gas removal processes can be *nonregenerative* or *regenerative.* The nonregenerative processes are suitable only when trace amounts of contaminants must be removed and/or very high purity of treated gas is desired. Nonregenerative processes become too costly when the H_2S to be removed exceeds about 1 ton per day. Examples of nonregenerable treating processes are SulfaTreat® and Chemsweet®, both marketed by C.E. Natco.

Regenerative processes are more economical for removing larger quantities of contaminants. An example of a regenerative process is the use of an aqueous amine solution to remove the H_2S and CO_2 from a sour gas stream. The amine solution is then regenerated by reducing its pressure and heating it to about 250°F. The solution is then cooled and recycled for reuse. Regenerative treating processes can be broadly classified as those that depend on:

- chemical reaction in
  - amine-based solvents,
  - nonamine based solvents;
- physical absorption;
- mixed chemical/physical absorption;
- adsorption on a solid.

CHEMICAL REACTION

In these processes, H_2S and/or CO_2 are chemically bound to the active ingredient in the treating solution. Therefore, the residue gas can be treated to retain only very low levels of these contaminants. The chemical solvent processes in current commercial processes use weak bases like *alkanolamines, alkali salt* solutions, *potassium carbonate,* or a *chelate* solution.

Amine-Based Solvents

These processes use an aqueous solution of a weak base that reacts chemically with the acid gas components in a gas stream. The reaction is usually reversible and takes place in a *contactor* vessel. The solution is regenerated by reversing the reaction in a regenerator by raising the solution's temperature and reducing its pressure. The regeneration energy requirements for chemical solvent processes are relatively high.

Aqueous alkanolamines, such as *diethanolamine (DEA)* and *methyldiethanolamine (MDEA)*, are the most widely used chemical solvents for the removal of acid gases from natural gas streams. These processes are particularly useful when low levels of acid gas components are desired in the treated gas. Another advantage of aqueous amine solutions is that they do not co-absorb much of the heavy hydrocarbon components present in the gas being treated (fig. 4.1).

In most gas treating applications, the purpose is to reduce hydrogen sulfide (H_2S) to 4 ppmv while reducing carbon dioxide (CO_2) to the sales gas specification of 2%–3%. This is done to remove less acid gases from the sour inlet gas, thereby reducing the treating requirements. This saves both operating and capital costs because of lower regeneration energy requirements and the reduction in the size of the plant's regeneration system.

Older acid gas removal processes were not selective and removed most of the CO_2 from the inlet gas along with the H_2S. Many of the current processes are more selective and, therefore, are used in newer engineering designs.

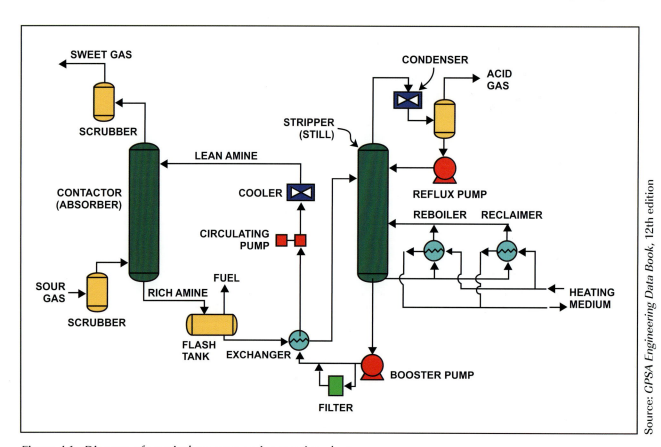

Figure 4.1 Diagram of a typical aqueous amine treating plant

Source: *GPSA Engineering Data Book*, 12th edition

Selective solvents are also replacing older treating solutions in existing plants because the payback on solvent replacement cost versus energy savings is usually less than 1–2 years. Also, replacing the older treating solvents with the selective solvents can provide additional treating capacity in an existing plant.

There are several different amine processes in commercial use today.

Monoethanolamine (MEA)

The use of *MEA* treating solvent has declined dramatically in recent years. The key reasons are the high regeneration heat requirements and high *corrosivity* of the MEA solution. The corrosion problem can be partially addressed by limiting solvent strength to 15%–20% by weight and keeping acid gas loadings below 0.3–0.4 mole per mole of amine. Processes, such as Amine Guard™ from Dow and GAS/SPEC® from Ineos LLC., market MEA processes that contain corrosion inhibitors.

MEA forms nonregenerable compounds with *cobalt (II) carbonyl sulfide (COS), and carbon disulfide (CS$_2$)* and, therefore, needs a *reclaimer* to remove these highly corrosive *degradation products*. Solution losses through *vaporization* from the contactor can be high because of MEA's relatively high vapor pressure. This problem can be minimized by using a water wash system.

Diethanolamine (DEA)

DEA is used for both gas and liquid hydrocarbon treating. The DEA process is more widely used than MEA because of its lower heat of regeneration and lower corrosivity. When compared to MEA, DEA's advantages are that:

- a DEA solution can remove more acid gases than MEA (0.35–0.65 mole of acid per mole of DEA versus 0.3–0.4 mole/mole of MEA);
- DEA forms regenerable products with COS and CS$_2$. A reclaimer is, therefore, not required;
- DEA can be used for the partial removal of COS and CS$_2$.

A key disadvantage of this process is that DEA is limited in its ability to *slip* or reject CO$_2$; that is, allowing part of the CO$_2$ to pass through the absorber without being absorbed into the treating solution. Also, DEA will not treat a gas to pipeline specifications that do not have as low an operating pressure as an MEA.

Diglycolamine® (DGA®)

Diglycolamine® (DGA®) is a primary amine capable of removing not only H$_2$S and CO$_2$ but also about 20%–50% of COS and mercaptans from hydrocarbon streams. DGA is used both in natural gas and refinery gas treating applications. It removes H$_2$S without a significant ability to slip CO$_2$. More than MEA and DEA, DGA has a higher affinity for aromatics and heavy hydrocarbons. Also, a DGA system does require a reclaimer to remove degradation products. High acid gas pickup can be achieved by using a 50%–70% by weight solution strength rather than MEAs 15%–20% by weight solution. Because of its low freezing point (–30°F for a 50% by weight solution), DGA has an advantage over other aqueous solutions for operation in colder climates.

Methyldiethanolamine (MDEA)

Aqueous methyldiethanolamine (MDEA)-based solvents have captured a large share of the current gas treating market. The main reasons for MDEA's widespread acceptance are its high selectivity for H_2S over CO_2 and lower regeneration heat requirements. Because of its relatively low vapor pressure, MDEA vaporization losses in a contactor are also lower than many other amines.

MDEA reacts rapidly with H_2S but much more slowly with CO_2. This fact is used to design an H_2S selective plant by limiting the contact time between the sour gas and MDEA solvent in a contactor. Some of the benefits of selective H_2S removal are:

- reduced circulation rates because of the ability to slip CO_2;
- higher H_2S in the acid gas, thereby providing a better feed to a *Claus process* to recover elemental sulfur.

Although MDEA reacts more slowly with CO_2 than with H_2S, it is an excellent solvent for the bulk removal of CO_2 from hydrocarbon streams (both vapor and liquid) when adequate reaction time is provided in the contactor. The regeneration heat requirements for MDEA solution are lower than many other amines. This combination of high acid gas loadings and low regeneration heat requirements results in an economical process for the bulk removal of CO_2 from gas streams.

Formulated Amines

There are many proprietary MDEA-based formulated solvents marketed by various companies. These formulated solvents retain the benefits of MDEA that are further enhanced with proprietary additives for specific applications. Some formulations allow slipping (rejection) of a larger percentage of the inlet gas CO_2 while reducing H_2S to pipeline specifications of 4 ppmv. At the other extreme, some formulations reduce CO_2 in the residue gas down to levels suitable for a liquefied natural gas (LNG) plant feed. Some benefits of formulated amines compared to primary, secondary, and tertiary MDEA are:

- reduced solution circulation rates;
- lower heat of regeneration; and
- smaller equipment due to lower circulation rates.

Some commercial examples of the MDEA-based formulated solvents are Shell's ADIP-X, Ineos LLC's GAS/SPEC®, Dow's Ucarsol™, and Huntsman's Textreat®. Other than ADIP-X, no process license fee is usually charged by formulated solvent suppliers. These companies sell the formulated amine solutions and in some cases, just the additives. For ADIP-X, Shell charges a licensing fee, but the customer can purchase the solvent components from the least expensive source.

Hindered Amines

Sterically hindered amine-based solvents were introduced more than 20 years ago. These amines have a molecular structure that makes them highly selective for H_2S removal in the presence of CO_2. Sterically hindered amines are most suitable for low-pressure refinery and *sulfur recovery unit (SRU)* tail-gas treating applications. Tail gas is the gas that remains after most of the sulfur has been removed in an SRU. An example of this technology is ExxonMobil's Flexsorb® solvents, which have somewhat higher licensing fees and solvents costs. However, many times the energy and equipment savings can offset these costs.

Table 4.1 and Table 4.2 provide approximate guidelines for several commercial amine processes and show the physical properties of many of the gas treating chemicals.

Table 4.1
Approximate Guidelines for Several Commercial Gas Processes

	MEA	DEA[8]	DGA®	Sulfinol	MDEA[8]
Acid gas pickup, scf/gal @ 100°F, normal range[1]	3.1–4.3	6.7–7.5	4.7–7.3	4-17	3-7.5
Acid gas pickup, mols/mol amine, normal range[2]	0.33–0.40	0.20-0.80	0.25-0.38	NA	0.20-0.80
Lean solution residual acid gas, mol/mol amine, normal range[3]	0.12±	0.01±	0.06±	NA	0.005-0.01
Rich solution acid gas loading, mol/mol amine, normal range[2]	0.45–0.52	0.21–0.81	0.35–0.44	NA	0.20–0.81
Solution concentration, wt%, normal range	15–25	30–40	50–60	3 comps., varies	40–50
Approximate reboiler heat duty, Btu/gal lean solution[4]	1,000–1,200	840–1,000	1,100–1,300	350–750	800–900
Steam-heated reboiler tube bundle, approx. average heat flux, Q/A—Btu/hr-ft² [5]	9,000–10,000	6,300–7,400	9,000–10,000	9,000–10,000	6,300–7,400
Direct-fired reboiler fire tube, average heat flux, Q/A—Btu/hr-ft² [5]	8,000–10,000	6,300–7,400	8,000–10,000	8,000–10,000	6,300–7,400
Reclaimer, steam bundle or fire tube, average heat flux, Q/A—Btu/hr-ft² [5]	6–9	NA[6]	6–8	NA	NA[6]
Reboiler temperature, normal operating range, °F[7]	225–260	230–260	250–270	230–280	230–270
Heats of reaction[9]; approximate: Btu/lb H_2S	610	720	674	NA	690
Btu/lb CO_2	660	945	850	NA	790

NA – not applicable or not available

NOTES:
1. Dependent upon acid gas partial pressures and solution concentrations.
2. Dependent upon acid gas partial pressures and corrosiveness of solution. Might be only 60% or less of value shown for corrosive systems.
3. Varies with stripper overhead reflux ratio. Low residual acid gas contents require more stripper trays and/or higher reflux rations yielding larger reboiler duties.
4. Varies with stripper overhead reflux ratios, rich solution feed temperature to stripper and reboiler temperature.
5. Maximum point heat flux can reach 20,000–25,000 Btu/hr-ft² at highest flame temperature at the inlet of a direct-fired fire tube. The most satisfactory design of fire tube heating elements employs a zone-by-zone calculation based on thermal efficiency desired and limiting the maximum tube wall temperature as required by the solution to prevent thermal degradation. The average heat flux, Q/A, is a result of these calculations.
6. Reclaimers are not used in DEA and MDEA systems.
7. Reboiler temperatures are dependent on the solution concentration flare/vent line back pressure and/or residual CO_2 content required. It is good practice to operate the reboiler at as low a temperature as possible.
8. According to total.
9. B.L. Crynes and R.N. Maddox, *Oil & Gas Journal*, p. 65-67, Dec. 15 (1969). The heats of reaction vary with acid gas loading and solution concentration. The values shown are average.

Source: *GPSA Engineering Data Book*, 12th edition

Table 4.2
Physical Properties of Gas Treating Chemicals

	Monoethanol-amine	Diethanol-amine	Triethanol-amine	Diglycol®-amine	Diisopropanol-amine	Selexol®
Formula	$HOC_2H_4NH_2$	$(HOC_2H_4)_2NH$	$(HOC_2H_4)_3N$	$H(OC_2H_4)_2NH_2$	$(HOC_3H_6)_2NH$	Polyethylene glycol derivative
Molecular Wt	61.08	105.14	148.19	105.14	133.19	280
Boiling point @ 760 mm Hg, °F	338.9	516.2 (decomposes)	680 (decomposes)	430	479.7	518
Freezing point, °F	50.9	82.4	72.3	9.5	107.6	−20
Critical constants						
Pressure, psia	868	474.7	355	547.11	546.8	—
Temperature, °F	662	827.8	957.7	756.6	750.6	—
Density @ 20°C, gm/cc.	1.018	1.095	1.124	1.058 @ 68°F	0.999 @ 30°C	1.031 @ 77°F
Weight, lb/gal	8.48 @ 60°F	9.09 @ 60°F	9.37 @ 68°F	8.82 @ 60°F		8.60 @ 77°F
Specific gravity 20°C/20°C	1.0179	1.0919 (30/20°C)	1.1258	1.0572	0.989 @ 45°C/20°C	—
Specific heat @ 60°F, Btu/lb/°F	0.608 @ 68°F	0.600	0.70	0.571	0.69 @ 30°C	0.49 @ 41°F
Thermal conductivity Btu/[(hr • sq ft • °F)/ft] @ 68°F	0.148	0.127	—	0.121	—	0.11 @ 77°F
Latent heat of vaporization, Btu/lb	180 @ 760 mmHg	288 @ 73 mmHg	230 @ 73 mmHg	219 @ 760 mmHg	185 @ 760 mmHg	—
Heat of reaction, Btu/lb of Acid Gas						
H_2S			−400	−674	—	−190 @ 77°F
CO_2			−630	−850	—	−160 @ 77°F
Viscosity, cp	24.1 @ 68°F	350 @ 68°F (at 90% wt. solution)	1013 @ 68°F (at 95% wt. solution)	40 @ 60°F	870 @ 86°F 198 @ 113°F 86 @ 129°F	5.8 @ 77°F
Refractive index, N_d 68°F	1.4539	1.4776	1.4852	1.4598	1.4542 @ 113°F	—
Flash point, COC, °F	200	298	365	260	255	304

	Propylene Carbonate	Methyldiethanol-amine	Sulfolane®	Methanol	10% Sodium Hydroxide
Formula	$C_3H_6CO_5$	$(HOC_2H_4)_2NCH_3$	$C_4H_8SO_2$	CH_3OH	
Molecular Wt	102.09	119.16	120.17	32.04	19.05
Boiling point @ 760 mm Hg, °F	467	477	545	148.1	217
Freezing point, °F	−56.6	−9.3	81.7	−143.8	14
Critical constants					
Pressure, psia	—		767.3	1153.9	
Temperature, °F	—		1013.8	464	
Density @ 20°C, gm/cc.	1.2057				
Weight, lb/gal		8.68	10.623 @ 30°C/30°C	9.254	
Specific gravity 20°C/20°C	1.293	1.0418	1.268	0.7917	1.110
Specific heat @ 60°F, Btu/lb/°F	0.335	0.535	0.35 @ 30°C	0.59 @ 5°-10°C	0.897
Thermal conductivity Btu/[(hr • sq ft • °F)/ft] @ 68°F	0.12 @ 50°F	0.159	0.114 @100°F	0.124	
Latent heat of vaporization, Btu/lb	208 @ 760 mm Hg	204	225.7 @212°F	474 @ 760 mmHg	
Heat of reaction, Btu/lb of Acid Gas					
H_2S	—				
CO_2					
Viscosity, cp	1.67 @ 100°F 19.4 cs @ −40°F 1.79 cs @ 100°F 0.827 cs @ 210°F	1.3 cs @ 50°F 0.68 cs @ 100°F 0.28 cs @ 212°F	10.3 @ 86°F 6.1 @ 122°F 2.5 @ 212°F 1.4 @ 302°F 0.97 @ 392°F	0.6 @ 68°F	1.83 @ 68°F 0.97 @ 122°F 0.40 @ 212°F
Refractive index, N_d 68°F	1.4209	1.469	1.481 @ 86°F	1.3286	
Flash point, COC, °F	270	265	350	58	

Source: *GPSA Engineering Data Book,* 12th edition

Nonamine-Based Processes

In these processes, either an alkaline salt or a *chelate* solution is used to remove the acid gas components.

Hot Potassium Carbonate Process

The hot potassium carbonate process uses an aqueous solution of potassium carbonate to remove acid gas components. The contactor and stripper both operate at temperatures of 230°F to 240°F. This process is not suitable for reducing H_2S and CO_2 down to pipeline specifications but is used mainly for the bulk removal of these components.

Redox Process

In a *redox process*, H_2S is directly converted to *elemental sulfur* and water by reacting with oxygen. *Oxidation* of H_2S is a slow process that is increased by the addition of an auxiliary redox reagent, which enhances the electron transfer process.

The two commercial iron redox processes are LO-CAT® and SulFerox®. Both processes use a water solution of iron that is held in solution by organic chelating agents. Iron in the solution oxidizes H_2S to elemental sulfur. The depleted solution is then circulated to an oxidizer and regenerated with air for reuse. SulFerox uses a much higher concentration of iron chelate solution compared to LO-CAT and, therefore, might have capital and operating costs.

Biochemical Processes

In this liquid phase of the *biological oxidation process*, H_2S is directly converted to elemental sulfur. First, H_2S reacts with a *caustic* solution containing bacteria to form sulfide. Then, the solution flows to a reactor section, where in the presence of air, *thiobacillus bacteria* convert the sulfides to elemental sulfur. The elemental sulfur is filtered from a side stream of the solution. It is marketable for use as a fertilizer, or it can be disposed of in a landfill as a nontoxic waste.

The Shell-Paques™ process is economical for up to about 30 tons per day of sulfur removal. Shell-Paques is used for the removal of H_2S from natural gas, amine regenerator off-gas, Claus tail gas residue, refinery fuel gases, and waste gases (figs. 4.2 and 4.3).

Figure 4.2 Schematic diagram of the Shell-Paques™ process

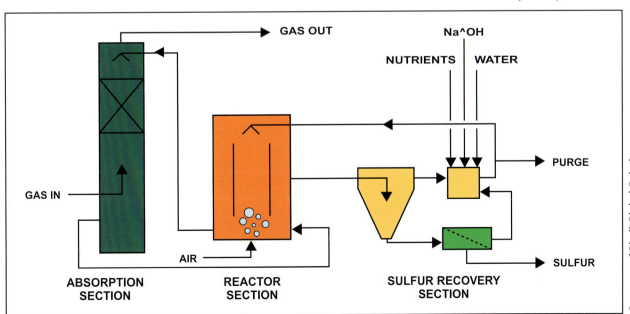

Figure 4.3   XTO Shell-Paques™ gas treating plant in Texas

It is especially suitable for gases containing oxygen that cannot be treated with other amines because of degradation and corrosion problems. The process can reduce H_2S in the treated gas to < 1 ppmv. This is greater than a 99.5% conversion of H_2S to elemental sulfur.

PHYSICAL ABSORPTION PROCESSES

In a physical absorption process, the solvent does not chemically react with the acid gas components. The physical solvent has a higher affinity for absorbing acid gases compared to lighter hydrocarbons, such as methane and ethane, which are the key components of natural gas. Therefore, a physical solvent preferentially absorbs acid gases contained in a hydrocarbon stream (fig. 4.4). In addition to H_2S, physical solvents can sometimes remove COS, CS_2, mercaptans, and water.

These processes normally operate at *ambient* or even subambient temperatures. High operating pressures are preferred; however, in some applications, low operating pressures (down to 15 psia) are possible.

Physical solvents are desirable when:
* the partial pressure of acid gas components in the feed is > 50 psi,
* the heavy hydrocarbon content in the feed gas is low;
* only bulk removal of acid gases is desired;
* selective removal of H_2S over CO_2 is desired.

If these conditions are met, physical solvent processes can be economical because relatively little energy is required to regenerate these solvents. The physical solvents are regenerated by either:
* multistage flashing to progressively lower pressures;
* regeneration at low temperature with an inert stripping gas; or
* heating and stripping of solution with steam/solvent vapors (reboiling).

Plant Processing of Natural Gas

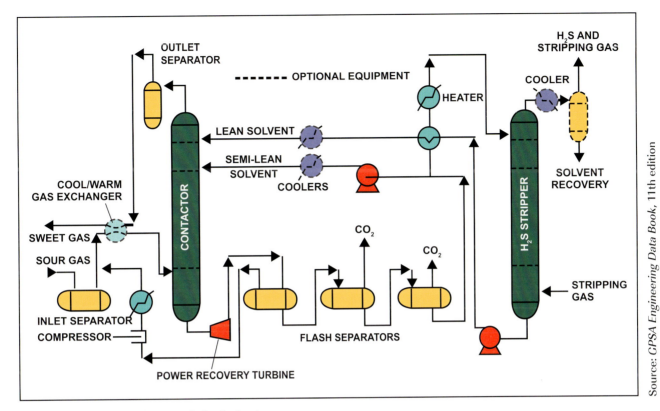

Figure 4.4   Diagram of a typical physical solvent process

There are several types of physical solvents marketed by various technology companies.

Selexol®

Marketed by Dow, Selexol® solvent has higher solubility for H_2S versus CO_2, making it possible to slip some CO_2 into the residue gas. In addition to H_2S and CO_2, the Selexol solvent can also remove some COS, CS_2, and mercaptans without degradation of the solvent. The solvent can also dehydrate natural gas if water is removed in the solvent regeneration step. The solvent capacity depends on the partial pressure of acid gases and increases as partial pressures increase. Loaded solvent is regenerated by flashing in one or more stages to successively lower pressures. To reduce hydrocarbon losses, the flash gas from one or more of the regeneration stages is compressed and recycled back to the contactor. The final flash stage might be under *vacuum*.

The disadvantages of the Selexol® process are:

- coabsorption of heavy hydrocarbons;
- the solvent is expansive;
- a user has to pay *royalty* charges.

Propylene Carbonate Process

The Fluor Solvent™ process uses propylene carbonate, a physical solvent that requires no heat for solvent regeneration. Propylene carbonate is nontoxic, nonfoaming, and biodegradable. It has a low affinity for hydrocarbons and requires no monitoring of solution chemistry. Because propylene carbonate is noncorrosive, carbon steel equipment can be used throughout the plant.

The advantage of the Fluor Solvent process over amine processes is its high acid gas loadings that result in a corresponding decrease in initial capital costs. Regeneration of the loaded solvent occurs with successive pressure reductions that require no regenerative heat. The disadvantages of this process are:

- coabsorption of heavy hydrocarbons;
- a higher vapor pressure than Selexol, therefore, higher solvent loses;
- solvent is thermally unstable;
- higher operating costs.

Rectisol® Process

Rectisol® is an acid gas removal process marketed by both Lurgi GmbH and Linde LLP. It uses methanol as a physical solvent to remove H_2S, CO_2, COS, etc., from synthesis gas, which is primarily hydrogen and carbon monoxide produced by the gasification of coal or heavy hydrocarbons. Because methanol has high vapor pressure at ambient temperature, the Rectisol process is operated at temperatures of –30°F to –100°F. With proper design, the process can achieve low levels of H_2S (1– 2 ppm), CO_2 (50–250 ppm), mercaptans, and COS (< 2 ppm).

MIXED CHEMICAL/PHYSICAL ABSORPTION

Mixed chemical/physical absorption processes use a mixture of an aqueous solution containing a chemical and a physical solvent. This combines the benefits of both chemical and physical solvent processes. The gases can be treated to high purity with high acid gas removal capacity, and the regeneration energy requirements are between those of chemical and physical solvent processes. Several companies market these types of processes.

Sulfinol® Process

Sulfinol®, marketed by the Shell Oil Company, uses an aqueous solution of Sulfolane™ with either *diisopropanolamine (DIPA)* or MDEA. Sulfinol with DIPA is referred to as Sulfinol-D, while that with MDEA instead is referred to as Sulfinol-M.

Sulfinol can have high loadings (moles of acid gas per mole of amine) at high acid gas partial pressures. Acid gas loadings of up to one mole per mole of amine are possible. The greatly reduced solution circulation rate of Sulfinol is the basis for its economic advantage. Other advantages of Sulfinol are:

- the ability to remove H_2S down to 1 ppmv and CO_2 to less than 50 ppmv;
- the ability to remove other sulfur compounds down to low levels (e.g., < 4 ppmv);
- that H_2S is selective (when using MDEA instead of DIPA);
- that it has low corrosivity and foaming tendencies.

The disadvantages are:

- a high affinity for heavy hydrocarbons, in particular aromatics; and
- DIPA in Sulfinol-D degrades in the presence of CO_2 and might require a solution reclaimer.

Other examples of mixed-solvent processes are Optisol™ (C-E Natco), Amisol® (Lurgi GmbH), and Flexsorb® PS (ExxonMobil).

ADSORPTION ON A SOLID

In these processes, a solid bed or surface is used to preferentially adsorb or accumulate acid gas components from a sour gas stream as it flows through the bed. The acid gas components are then stripped from the bed by sending a high-temperature regeneration gas stream through the bed.

Molecular Sieve Process

The molecular sieve process involves the adsorption of acid gases on molecular sieves (fig. 4.5). It can be designed to remove H_2S and some CO_2 or to remove both H_2S and CO_2 along with COS, CS_2, and mercaptans. The operation of a molecular sieve unit is limited to gases with low acid gas content ($\leq$ 15 ppmv of H_2S) and/or low gas volumes ($<$ 10 MMscf/day). Molecular sieve units are often used to remove H_2S, CO_2, and other sulfur compounds. They also can be used to dehydrate gas before it goes to a natural gas liquefaction plant. In this process, H_2S can be removed to meet 4 ppmv pipeline specifications. The cycle times of a sieve bed are generally about 6 to 8 hours. The sieve regenerates for approximately 1 hour or more at a temperature of 450°F–550°F in a heat soak that removes all adsorbed materials.

The regeneration of a molecular sieve bed concentrates the H_2S into a small regeneration stream, which must be disposed of or treated. During the regeneration cycle, H_2S will exhibit a peak concentration in the regeneration gas of about 30 times the concentration of H_2S in the inlet stream. Understanding the concentration of this stream is essential when designing a gas treater for the regenerated gas.

During the sulfur peak of the operation, the regenerated gas is flared in small units. During the rest of the regeneration cycle, the regeneration gas stream is recycled to the feed stream. Operation of a molecular sieve plant is simple, but the design is complex. Molecular sieves are produced by several companies, including UOP LLC.

Figure 4.5   An integrated natural gas desulfurization plant

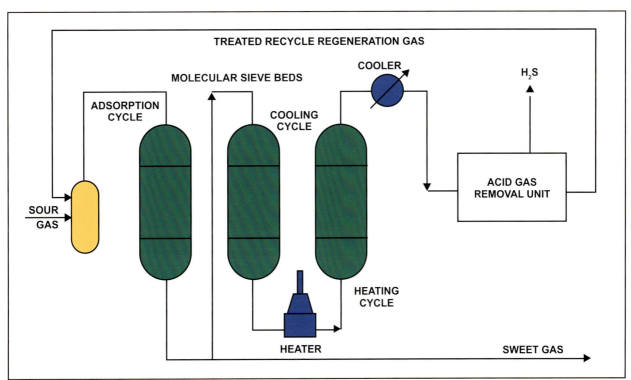

Source: *GPSA Engineering Data Book*, 12th edition

Activated Carbon Process

When using an activated carbon process, H_2S and other sulfur compounds are physically adsorbed, and a treated gas purity of 0–0.2 ppmv sulfur can be obtained. The adsorbed sulfur compounds are desorbed when regenerating the activated carbon with steam. This process is attractive for gases containing less than about 30 ppmv H_2S.

MEMBRANE PROCESSES

In *membrane processes, gas permeation* through a polymeric film or membrane is used to separate the acid gas components. Gas permeation is the transport of gas molecules through a thin polymeric film from a region of high pressure to one of low pressure. It is based on the principle that some gases are more soluble in polymeric membranes and pass more readily through them than other gases. While the acid gas components can readily permeate the membrane, the hydrocarbons permeate to a lesser degree. For example, CO_2, H_2S, and water are highly permeable compared to methane, ethane, N_2, and other hydrocarbons. Therefore, CO_2 and H_2S can readily be separated from a sour gas stream in a membrane process.

As the sour gas passes over a polymeric membrane at high pressure, CO_2 and H_2S are separated from the main gas stream into a low-pressure permeate stream. The rate at which CO_2 and H_2S permeate the membrane depends on the partial pressure difference between the feed gas and the permeate side of the membrane, the ratio of feed-to-permeate pressures, and the operating temperature of the membrane.

The gas-separation membranes are manufactured either as flat sheets or as hollow fibers. The sheets are typically combined into a spiral-wound element, while the hollow fibers are combined into a small bundle similar to a shell and tube exchanger.

In the spiral wound arrangement, two flat sheets of the membrane with a permeate spacer in between are glued along three sides to form an envelope or leaf that is open on one side (fig. 4.6).

Many of these leaves are separated by feed spacers and wrapped around a permeate tube with the open end facing the permeate tube. Sour feed gas containing CO_2 enters along the side of the membrane and passes through the feed spacers separating the leaf. As the gas travels between the leaves,

Figure 4.6 UOP's spiral wound membrane element

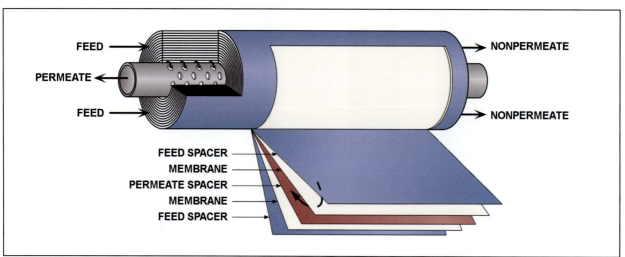

FEED → NONPERMEATE
PERMEATE ←
FEED → NONPERMEATE

FEED SPACER
MEMBRANE
PERMEATE SPACER
MEMBRANE
FEED SPACER

Courtesy of UOP, a Honeywell Company

Figure 4.7 Element being inserted into the casing

CO$_2$ and H$_2$S permeate the envelope. These permeated components have only one outlet; they must travel within the leaf and into the permeate tube. The transport is driven by the pressure difference between the feed gas and the permeate stream. The permeated gas enters through holes in the tube and travels down to join permeate from other tubes. The gas that does not permeate exits through the side of the element of the feed position (fig. 4.7).

The spiral-wound or hollow-fiber elements are housed in pressure tubes with up to six elements per tube (fig. 4.8).

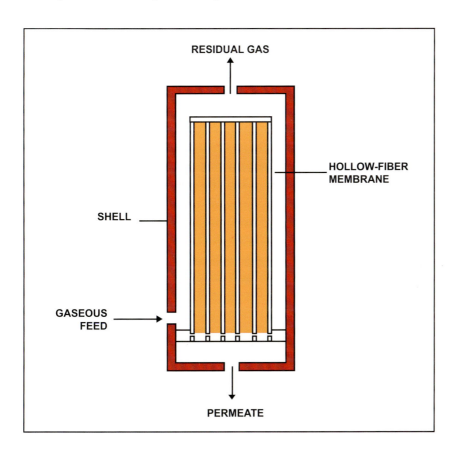

Figure 4.8 A hollow-fiber membrane element

These membrane systems are modular, skid-mounted units that contain either spiral-wound or hollow-fiber membrane elements (fig. 4.9). Based on the volume of gas being treated and the degree of treating required, these elements are arranged either in series or in parallel on the skid.

Membrane element plants have been used to remove CO_2 from 5 to 500 MMscfd of natural gas. Normally, the operating pressures range from 400 to 1250+ psig with CO_2 levels from 3 to over 60% in the feed gas. Some advantages of membranes include:

- lower capital and operating costs;
- can dehydrate and remove CO_2 to pipeline specs;
- smaller than an amine plant footprint;
- can be used on offshore platforms;
- can be located in remote areas;
- faster startup and shutdown.

Some disadvantages are:

- economical only for bulk removal of CO_2;
- significant loss of hydrocarbons (2%–3% of the inlet) in the permeate stream;
- membranes can easily be damaged by free hydrocarbon liquid and particulates.

Extensive gas pretreatment is usually required before it is fed to a membrane plant. The sour gas is first filtered to remove any particulates and aerosols. The gas is then heated to maintain a constant operating temperature and to prevent hydrocarbon condensation. The rigorous and extensive pretreatment processes might increase the costs of a membrane plant, making it (in some cases) less competitive compared to more conventional technologies. Two examples of commercially available membranes for CO_2 removal are Natco's Cynara® and UOP's Separex™.

Courtesy of UOP, a Honeywell Company

Figure 4.9   A membrane skid for the removal of CO_2 from natural gas

Plant Processing of Natural Gas

GENERAL OPERATING CONSIDERATIONS FOR GAS TREATING

When designing a reliable and efficient treating plant operation, all possible contaminants in the inlet stream must be considered. The key areas requiring attention are discussed next.

Inlet Separation

Inadequate inlet separation can cause severe problems for both acid gas absorption and downstream regeneration equipment. The inlet separation vessels should have adequate surge capacity to handle slugs of liquid hydrocarbons, water, and/or well treatment chemicals. In cases where solids or liquids are expected, high-efficiency filtration should be installed.

Foaming

Solution foaming causes poor contact between gas and the solvent, thereby reducing plant capacity and treating efficiency. Severe foaming is indicated when there is a sudden increase in pressure drop across a contactor. If not controlled, severe foaming can cause carryover of the treating solvent into the residue gas. Some of the causes for foaming are:

- suspended solids;
- presence of organic acids in the solvent;
- condensed hydrocarbons;
- corrosion inhibitors entrained in the inlet gas;
- degradation products.

Temporary foaming incidents can be controlled by the injection of antifoam chemicals into the circulating solvent. Antifoam agents are either silicon or alcohol based.

Filtration

To prevent solution foaming, it is necessary to filter 10%–100% of the circulating solution to remove any entrained solids. In most cases, cartridge filters of 5 microns are recommended on the rich solution following the flash step. An activated carbon filter is used on the lean circulating solution to remove any heavy hydrocarbons or other chemicals that might promote foaming. Cartridge filters can also be added after the carbon filter to prevent carbon fines from building up in the solution.

Corrosion

All treating equipment experiences corrosion when H_2S and/or CO_2 are present along with water. Both of these components form weak acids with water that makes them corrosive. CO_2 alone, or with a small amount of H_2S, is more corrosive. The reason is that H_2S reacts with iron to form an iron-sulfide film that adheres to the metal to provide some protection against further corrosion. Amine-degradation products can also contribute to corrosion. Ingress of oxygen into a treating solution must be avoided at all costs, as it contributes to degradation of amines and the formation of corrosive organic acids.

REFERENCES

Dortmundt, D., Doshi, K. "Recent Developments in CO_2 Removal Technology," UOP LLC, Des Plains, Illinois, http://www.uop.com/objects/84CO2RemvbyMembrn.pdf (1999).

GPSA Engineering Data Book, 11th edition, Gas Processors Suppliers Association, Tulsa, Oklahoma (2004).

UOP LLC, "Gas Processing" UOP Web site, http://www.uop.com/gasprocessing/6000.html.

SULFUR RECOVERY

Gas treating plants must strictly comply with legal, government, and safety standards and regulations concerning emissions and pollution. During the treatment process, H_2S and some or most of the CO_2 are removed from the sour gas stream, as discussed in Chapter 4. These removed sour gas components must be dealt with cautiously.

While the emissions requirements vary with geography, most countries do not permit the emission of more than a few pounds of sulfur (H_2S, SO_2, etc.) per day into the atmosphere. To control emissions, the acid gas stream from a treating plant is fed to a sulfur recovery unit (SRU) where H_2S and other sulfur compounds are converted and recovered as nontoxic elemental sulfur (S). The tail gas from the SRU still contains some sulfur components. These are converted to SO_2 in an incinerator before being discharged into the atmosphere. If high (99.8+%) sulfur recovery is desired, the SRU tail gas is fed to a tail-gas treating plant for further reduction of sulfur emissions.

Thermal Process

In 1883, an English scientist, Carl Friedrich Claus, discovered and patented a process in which H_2S was reduced to elemental sulfur and water in the presence of a catalyst.

Claus's formula for this process is:

$$H_2S + \tfrac{1}{2} O_2 = S + H_2O$$

The control of this *exothermic*, or heat-releasing, process was difficult, and conversion to elemental S was low. The modified Claus process used today overcomes the control and conversion problems by dividing the Claus process into the following two steps:

Thermal Step

In this exothermic step, the air-to-acid gas ratio is controlled so that about ⅓ of the H_2S is oxidized to SO_2. For gases containing hydrocarbons and/or ammonia from a sour water stripper, enough air is injected to ensure complete combustion of ammonia and hydrocarbon components. The process temperature during this step is from 1,800°F–2,500°F.

Catalytic Step

In this moderately exothermic catalytic reaction, the sulfur dioxide (SO_2) formed in the thermal section reacts with unburned H_2S to form gaseous elemental sulfur. The catalyst used in this step is made from activated alumina or titanium dioxide.

Other than the reactions of hydrocarbons and other combustibles, the key reactions taking place are:

Thermal Reaction:

$$H_2S + 1\tfrac{1}{2} O_2 = SO_2 + H_2O$$

Thermal and Catalytic Reaction:

$$2 H_2S + SO_2 = 3 S + 2 H_2O$$

5
Sulfur Recovery and Claus Off-Gas Treating

The Overall Reaction:

$$3 \, H_2S + 1\frac{1}{2} \, O_2 \ = \ 3 \, S + 3 \, H_2O$$

This is a simplified interpretation of the reactions taking place in a Claus unit. The actual reaction equilibrium is complicated by the presence of various gaseous elemental sulfur species or forms that exist at different temperatures. These species are S_2, S_3, S_4, S_5, S_6, S_7, and S_8. Figure 5.1 shows some of the different sulfur types that exist at different temperatures.

In addition, the side reactions of hydrocarbons, H_2S, and CO_2 present in the acid gas feed can cause the formation of carbonyl sulfide (COS), carbon disulfide (CS_2), carbon monoxide (CO), and hydrogen (H_2) [2,3].

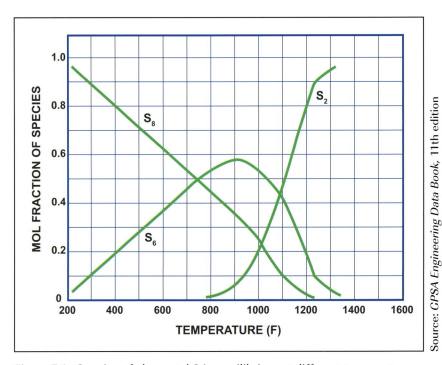

Source: *GPSA Engineering Data Book*, 11th edition

Figure 5.1   Species of elemental S in equilibrium at different temperatures

Catalytic Recovery

Following the thermal reaction furnace and waste heat recovery units, the catalytic recovery of sulfur consists of three substeps: heating, catalytic reaction, and cooling plus condensation. These three steps are usually repeated a maximum of three times (fig. 5.2).

When a tail gas treatment unit is added downstream of the Claus plant, only two catalytic stages are normally installed. Figure 5.3 is a mechanical arrangement of a small, packaged two-stage Claus plant where the tail gas is sent to an incinerator.

Plant Processing of Natural Gas

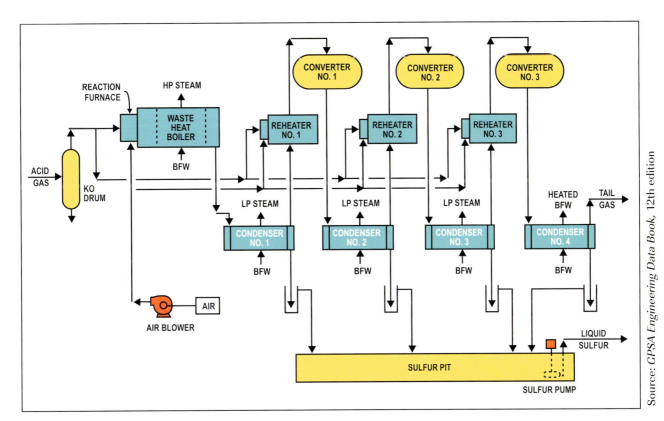

Figure 5.2   Three-stage modified Claus sulfur-recovery unit

Source: GPSA Engineering Data Book, 12th edition

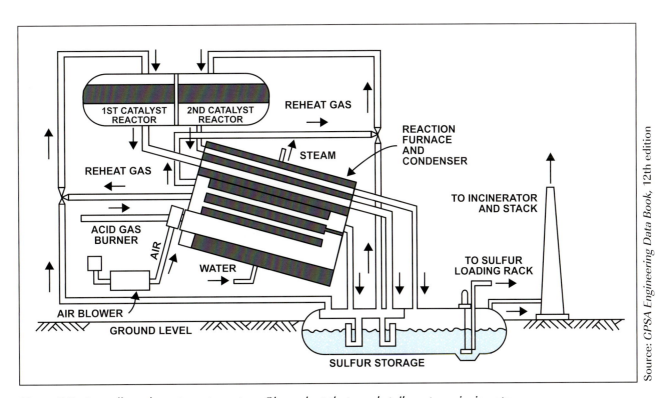

Figure 5.3   A small, package-type, two-stage Claus plant that sends tail gas to an incinerator

Source: GPSA Engineering Data Book, 12th edition

Sulfur Recovery and Claus Off-Gas Treating

The first substep in a catalytic recovery stage is the gas heating process. Sulfur condensation in the catalyst bed must be prevented because it can lead to catalyst fouling. The necessary bed operating temperature for the individual stages is achieved by heating the process gas in a reheater. The typical operating temperature of the first catalyst stage is 600°F–630°F. The high temperature in the first stage is needed to *hydrolyze,* or convert, COS and CS_2 to CO_2 and H_2S in the presence of water. COS and CS_2 are formed in the thermal step and would not otherwise be converted to SO_2 in the Claus process. The catalytic conversion is maximized at lower temperatures, but care must be taken to ensure that each bed is operated above the dew point of sulfur. The operating temperatures are typically 460°F for the second stage and 390°F for the third stage.

In the sulfur condensers, the gas coming from the catalytic reactors is cooled to 265°F–300°F by generating low-pressure steam on the shell side of a condenser. The liquid sulfur streams from the sulfur condensers are routed to a subsurface *sulfur pit.* The tail gas from the Claus unit containing sulfur compounds (H_2S, SO_2, COS, CS_2, etc.), H_2O, H_2, CO, and N_2 is either burned in an incineration unit or further desulfurized in a downstream tail gas treatment unit.

CLAUS OFF-GAS TREATING

Although the Claus process can convert more than 96% of the sulfur compounds to elemental sulfur, the off-gas from the Claus unit still contains more sulfur than atmospheric emission regulations allow. In the United States, depending on the amount of sulfur in the sour gas, over 99.9% sulfur recovery might be required. This is achieved by sending the Claus off-gas to the tail-gas treating units.

SCOT Process

A process developed to achieve very low sulfur emissions is the *Shell Claus Off-Gas Treating (SCOT)* process (fig. 5.4). The SCOT process is reliable and can increase overall sulfur recovery to > 99.8+%. It can reduce the final vent gas to <10ppmv H_2S by using the low sulfur SCOT process. The SCOT process has been used for SRU capacity range of 3–4,000 tons per day.

In the SCOT process, all non-H_2S sulfur species (CS_2, mercaptans, elemental sulfur, SO_2) are converted to H_2S in the reactor in the presence of a reducing gas-like hydrogen. The off-gases from the reactor are cooled, first in a waste heat boiler, and further cooled in a *quench column* or *tower.* The quench column overhead gas, containing H_2S, is then sent to an amine unit for H_2S recovery. The acid gases, from the amine unit's regenerator, are recycled back to the Claus unit inlet for reprocessing.

The residue gas from the SCOT amine absorber is sent to an incinerator to convert any residual sulfur to SO_2 before being vented to the atmosphere. There are nearly 200 SCOT units in operation worldwide (fig. 5.5).

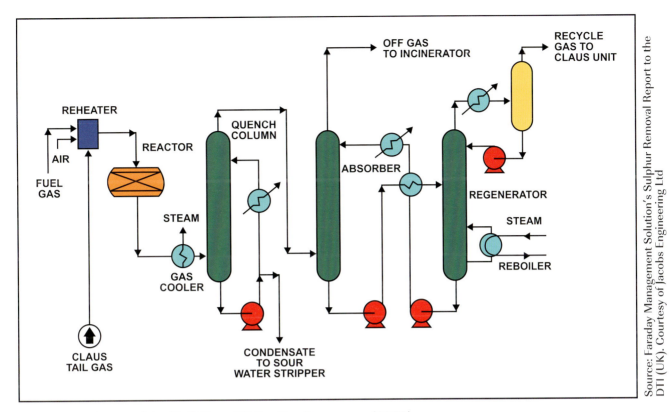

Source: Faraday Management Solution's Sulphur Removal Report to the DTI (UK). Courtesy of Jacobs Engineering Ltd

Figure 5.4   Diagram of the Shell Claus Off-Gas Treating process (SCOT)

Courtesy of Shell Global Solutions

Figure 5.5   SCOT process plant

Sulfur Recovery and Claus Off-Gas Treating

REFERENCES

Gamson, B.W., Elkins, R.H., "Sulfur from Hydrogen Sulfide," AIChE Chemical Engineering Progress (April 1953), Wiley Press, New York, New York.

Gary, J.H., Handwerk, G.E., "Petroleum Refining Technology and Economics," 2nd edition, Marcel Dekker, Inc., New York, New York (1984).

GPSA Engineering Data Book, 11th edition, Gas Processors Suppliers Association, Tulsa, Oklahoma (2004).

GPSA Engineering Data Book, 12th edition, Gas Processors Suppliers Association, Tulsa, Oklahoma (2004).

Kohl, A.L., Nielsen, R.B., "Gas Purification," 5th edition, Gulf Publishing Company, Houston, Texas (1997).

DEHYDRATION

All hydrocarbon fluids can retain some water. Water is soluble in liquid hydrocarbons and can be held in the vapor phase by hydrocarbon gases. When a liquid or gas is cooled, its capacity for containing water decreases. As a result, it can produce liquid and/or solid water called hydrates. Hydrates are a separate and problematic part of gas processing.

Figure 6.1 shows how the water content of natural gas varies with temperature and pressure. For example, at 1,000 psia and 100°F, water-saturated natural gas contains about 62 pounds of water per million standard cubic feet (MMscf) of gas. At 1,000 psia and 0°F, the gas contains only about 2 pounds of water per MMscf of gas.

The water dew point of a gas or liquid is the temperature at which free water will begin to separate from the gas or liquid. If a natural gas stream at 1,000 psia contains 62 pounds of water per MMscf of gas, its water dew point is 100°F. If it is cooled below 100°F, free water will be present.

Hydrates will form if a gas or liquid containing free water is cooled below its hydrate temperature. The graph shown in figure 6.2 can be used to estimate the hydrate temperature of natural gas. For instance, a 0.6 specific gravity gas has a hydrate temperature of about 61°F at 1,000 psia. If this gas must be cooled below its hydrate temperature, either due to pipeline transportation, pressure reduction for consumption, or for processing, precautions must be taken to prevent free-water dropout that causes freezing and formation of hydrates.

Hydrates are solid compounds that form as crystals and resemble snow. They are created by a reaction of natural gas with water, and when formed, are about 10% hydrocarbon and 90% water. Hydrates have a specific gravity of about 0.98 and will float in water and sink in hydrocarbon liquids. Freezing can be avoided by either removing the water from the gas or liquid prior to cooling below the hydrate temperature or by using a hydrate inhibitor to mix with the water condensed during cooling.

Dehydration is the process of removing water from a substance. Dehydration can be accomplished by using solid substances such as those used in dry-bed dehydrators. It can also be done using a liquid, such as triethylene glycol. A *stripping gas* in a TEG reboiler can also be used. The more common hydrate inhibitors are methanol and ethylene glycol. Most natural gas facilities use one or more of the following dehydration processes: ethylene glycol injection, TEG dehydration, or dry-bed dehydrators.

6
Dehydration and Mercury Removal

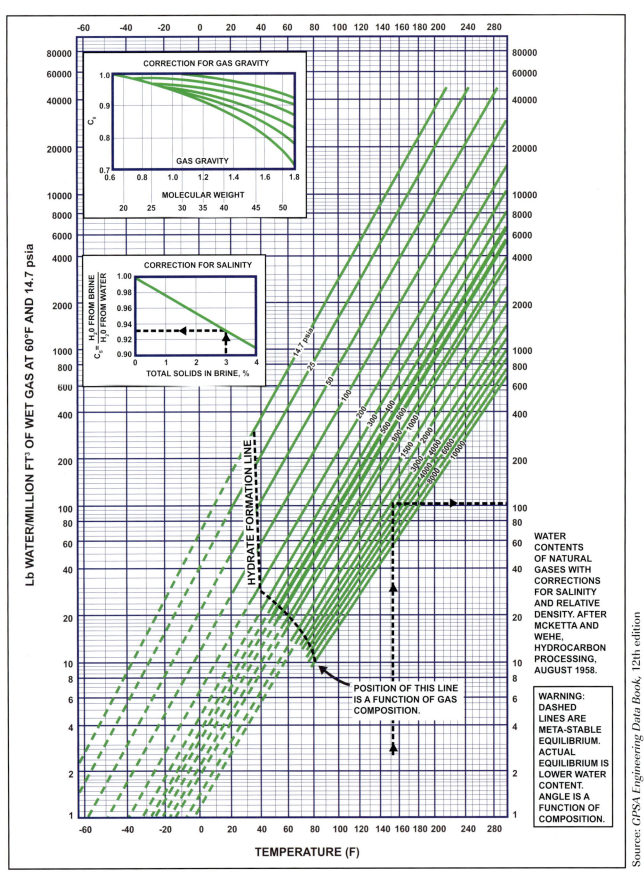

Figure 6.1   Water content of natural gas varies.

Source: *GPSA Engineering Data Book*, 12th edition

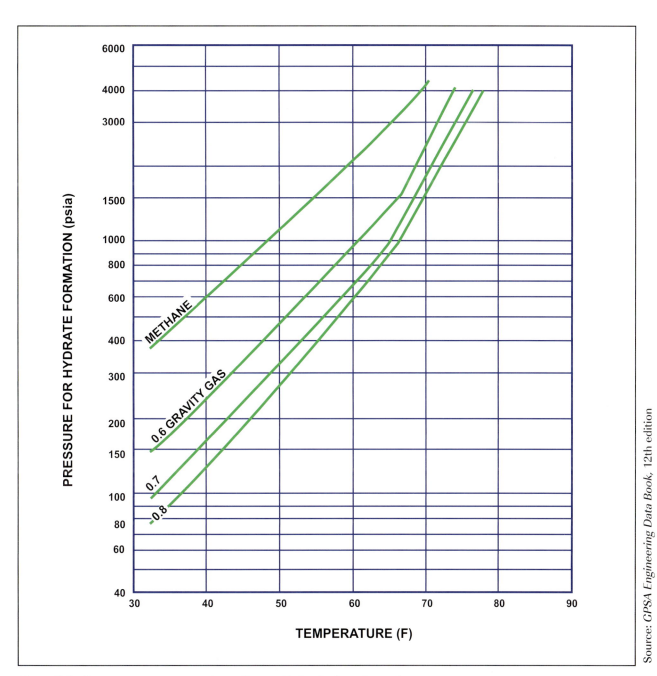

Figure 6.2 Pressure-temperature curves for predicting hydrate formation

Inhibitor Injection

Chemical inhibitors are commonly injected into natural gas to prevent hydrate formation. Thermodynamic and *low dosage hydrate inhibitor (LDHI)* are the most common types of hydrate inhibitors. *Thermodynamic inhibitors* lower the temperature of hydrate formation. Thermodynamic inhibitors include ethylene glycol (EG), diethylene glycol (DEG), triethylene glycol (TEG), and methanol. LDHIs include both *kinetic hydrate inhibitors (KHIs)* and *antiagglomerants (AAs)*, which reduce the effect of formation in a different way. KHIs lower the rate of formation for a specific length of time, while AAs limit the size hydrate crystals to a fraction of a millimetre. Figure 6.3 shows the physical properties of the glycols and methanol used as inhibitors.

Prior to cooling the gas to its hydrate-forming temperature, an inhibitor must be injected upstream. Thorough mixing of the inhibitor with the gas is required, especially in all parts of chillers and heat exchangers. The effectiveness of inhibitors is a measure of the hydrate point depression needed for the lowest operating temperature. In other words, the injected inhibitor provides the method to prevent hydrate formation.

	ETHYLENE GLYCOL	DIETHYLENE GLYCOL	TRIETHYLENE GLYCOL	TETRAETHYLENE GLYCOL	METHANOL
Formula	$C_2H_6O_2$	$C_4H_{10}O_3$	$C_6H_{14}O_4$	$C_6H_{18}O_5$	CH_3OH
Molecular Weight	62.1	106.1	150.2	194.2	32.04
Boiling Point* at 760 mm Hg, °F	387.1	472.6	545.9	597.2	148.1
Boiling Point* at 760 mm Hg, °C	197.3	244.8	285.5	314	64.5
Vapor Pressure at 77°F (25°C) mm Hg	0.12	<0.01	<0.01	<0.01	120
Density (g/cc) at 77°F (25°C)	1.110	1.113	1.119	1.120	0.790
(g/cc) at 140°F (60°C)	1.085	1.088	1.092	1.092	-
Pounds per Gallon at 77°F (25°C)	9.26	9.29	9.34	9.34	6.59
Freezing Point, °F	8	17	19	22	-144.0
Pour Point, °F	-	-65	-73	-42	
Viscosity in Centipoise at 77°F (25°C)	16.5	28.2	37.3	44.6	0.52
at 140°F (60°C)	4.68	6.99	8.77	10.2	-
Surface Tension at 77°F (25°C), dynes/cm	47	44	45	45	22.5
Refractive Index at 77°F (25°C)	1.430	1.446	1.454	1.457	0.328
Specific Heat at 77°F (25°C) Btu/(lb•°F)	0.58	0.55	0.53	0.52	0.60
Flash Point, °F (PMCC)	240	255	350	400	53.6
Fire Point, °F (C.O.C.)	245	290	330	375	-

*Glycols decompose at temperatures below their atmospheric boiling point.
Approximate decompostion temperatures are:
Ethylene Glycol: 329°F   Triethylene Glycol: 404°F

Figure 6.3   Physical properties of selected glycols and methanol

Source: *GPSA Engineering Data Book*, 12th edition

Plant Processing of Natural Gas

Methanol is a commonly selected inhibitor due to its lower freezing point temperature when compared to glycols. An 85% aqueous solution of methanol has a *eutectic freezing point* of –200°F. The eutectic freezing point of a liquid is the lowest temperature at which the liquid can exist. This lower freezing temperature allows methanol to be used for expander plant hydrate control. Regeneration for methanol might not be financially necessary. Methanol is a consumable commodity and its cost is usually recovered in sales. However, problems can occur in natural gas liquid (NGL) recovery operations due to methanol's high vapor pressure and compatibility with NGL recovery processes downstream. Many of these recovery processes operate at temperatures lower than –200°F.

Ethylene glycol (EG) is the one of the most popular injection inhibitors because of its low cost, low viscosity, and low solubility in liquid hydrocarbons. Figure 6.4 depicts a typical EG injection system for a refrigerated lean-oil gas plant in which the concentrated or lean EG is injected upstream of the gas/gas exchanger and chiller.

To prevent freezing, EG mixes with condensed water as the gas cools. The glycol-water mixture or *rich glycol* is separated from the hydrocarbon liquid and vapor in the glycol separator. The glycol is then warmed in the exchanger and fed to the *flash tank* where the pressure is reduced to about 60 psig. Heat and the reduction of pressure reject the light hydrocarbons (primarily methane) that are absorbed. Glycol is regenerated for reconcentration in the reboiler where heat vaporizes the water. The *lean glycol*, after leaving the surge tank and passing through the exchanger, is reinjected into the inlet gas.

Figure 6.4 A typical EG injection system

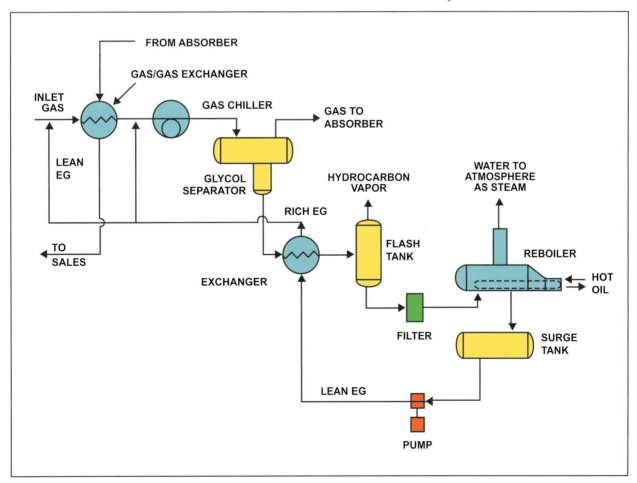

Problems in the glycol injection systems can be divided into three categories:

- Glycol concentration
- Glycol circulation
- Glycol solution condition

The requisite concentration of glycol must be produced to prevent freezing and avoid losses from regeneration. As shown in figure 6.5, the maximum concentration of lean glycol is 80% for effective performance. It is essential to know the amount of water condensation to determine the rich glycol concentration because this ultimately affects hydrate formation and reboiler temperatures.

In figure 6.6, the reboiler temperatures are given for EG at different concentrations. Note on this graph that temperature depends on the plant's geographic elevation.

An adequate lean glycol circulation rate is required to provide adequate mixing with the inlet gas. The desired concentration of rich glycol depends on the lean glycol concentration, the circulation rate, and the amount of water condensed. The rich glycol has a specific concentration to produce the desired hydrate-depressing property and prevent freezing. The hydrate depression is the difference between the hydrate forming temperature of the gas minus the desired temperature achieved using an inhibitor.

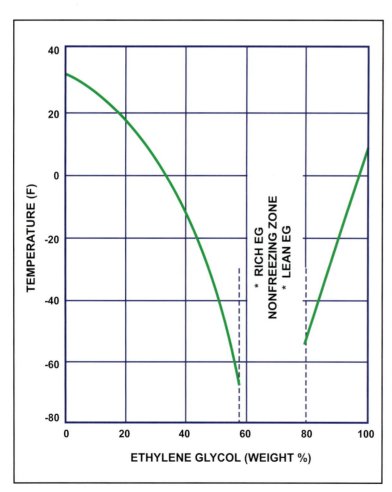

Figure 6.5 Freezing temperatures of ethylene glycol-water mixtures

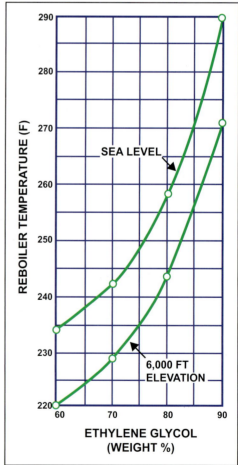

Figure 6.6 Glycol reboiler temperatures

Plant Processing of Natural Gas

The lowest operating temperature should always be above the freezing temperature of the rich glycol solution. Having calculated the hydrate depression for operation, figure 6.7 can be used to determine the concentration of rich glycol needed.

The condition of the glycol solution is determined by the amount of solid material and heavy hydrocarbons present. These solids must be continuously filtered from the circulating glycol stream. Heavy hydrocarbon liquids must be removed from the rich glycol solution in the glycol separator to prevent foaming and glycol loss. Adequate separation depends mainly on residence time and temperature. Longer residence time and higher temperature results in better separation. If separation efficiency cannot be improved, a charcoal filter can be used to remove liquid hydrocarbons and prevent their buildup and foaming.

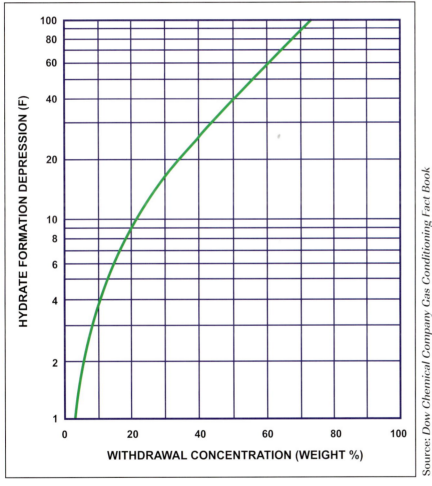

Source: Dow Chemical Company Gas Conditioning Fact Book

Figure 6.7   Hydrate depression versus minimum withdrawal concentration of ethylene glycol

Example Problem:

Assume a 0.6 specific gravity gas at 1,000 psia is cooled to 0°F. Determine the hydrate forming temperature, the rich ethylene glycol concentration needed, and the reboiler temperature required at sea level. Verify that this glycol solution will not freeze at operating conditions.

Solution:

Referring to figure 6.2, it can be seen that the gas has a hydrate temperature of 61°F. With a required hydrate depression of 61°F (61°F-0°F), figure 6.7 shows that a 62%-rich ethylene glycol concentration is needed. The reboiler temperature at sea level required to reconcentrate the glycol is about 235°F, as shown in figure 6.6. Figure 6.5 can be used to verify that a 62%-rich glycol solution will prevent freezing at 0°F.

DEHYDRATION METHODS

Dehydration should be used in situations where inhibitors are not economical or practical. Selection of a dehydration process depends on the desired level of dehydration, compatibility with connected process operations, economics, and experience. There are several commonly used dehydration processes.

Liquid Desiccants

When inhibition to get the desired dew-point depression (typically 60°F–120°F with glycols) is impractical, more economical liquid desiccants are used instead of solid desiccants. An *absorber contactor dehydration system*, or glycol system, can use various glycols to absorb water in a contactor. After regeneration using heat from a reboiler, the glycol is recirculated to the absorber or contactor. TEG is the most commonly used glycol because it can be regenerated to a high concentration without major degradation. DEG and *tetraethylene glycol (TREG)* are also used in some applications. The TEG dehydration system is illustrated in figures 6.8a and 6.8b. The process can be easily automated for unattended operation.

An inlet *gas scrubber*, located upstream of the contactor or absorber, is used to remove any accidental carryover of water, hydrocarbons, treating chemicals, or corrosion inhibitors into the contactor. Even small quantities of these contaminants can cause glycol loss due to foaming, reduced efficiency, and increased maintenance.

The process begins as lean TEG flows down the contactor countercurrent to the rising inlet gas, absorbing water and leaving the bottom of the contactor tower rich with water. The absorber uses bubble-cap trays or *structure packing* to allow for better contact between liquid and gas. Structure packing allows for a smaller column diameter and height.

The rich TEG is reduced in pressure at the flash tank where light hydrocarbons, mainly methane, can escape from the solution and leave the top as vapor. Flash-tank pressures are generally less than 75 psia and require typical resident time of 2–5 minutes for degassing and 20–30 minutes for sufficient liquid hydrocarbon separation. After heat exchange and solid filtration, the rich TEG enters the reboiler where heat distills water from the solution and exits the top as vapor.

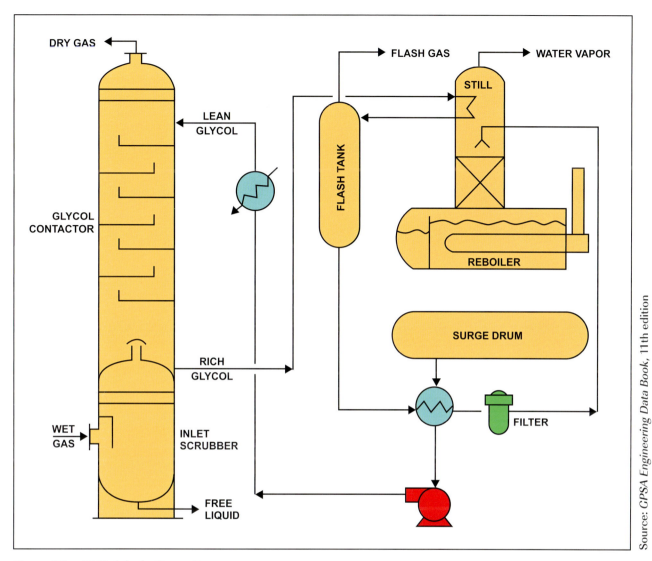

DRY GAS

LEAN
GLYCOL

GLYCOL
CONTACTOR

FLASH GAS

WATER VAPOR

STILL

FLASH TANK

RICH
GLYCOL

REBOILER

SURGE DRUM

WET
GAS

INLET
SCRUBBER

FREE
LIQUID

FILTER

Source: GPSA Engineering Data Book, 11th edition

Figure 6.8a   TEG dehydration unit

*Figure 6.8b   Glycol regeneration
unit designed to regenerate TEG
for natural gas dehydration on an
offshore oil and gas production
platform*

Dehydration and Mercury Removal

81

Most regenerators operate at about 400°F. At temperatures higher than 400°F, the TEG starts to degrade. The reboiler can either be heated by steam, hot oil, or a direct-fired heater. The TEG concentration increases in the glycol regenerator. It is then cooled and recirculated to the contactor tower with low water content. This regeneration happens at or near atmospheric pressure and produces a 98.6%–99.0% TEG concentration with a corresponding inlet gas dew-point depression of 100°F–104°F. The lower the desired water content in the lean gas, the higher the TEG concentration required to achieve greater dehydration of the inlet gas.

Lowering the H_2O partial pressure can achieve TEG purity >98.6% in the regenerator. Higher TEG concentrations produce greater dew-point depression of the inlet gas. The most common methods use a stripping gas or vacuum in the regenerator. Operation of the regenerator under a vacuum lowers the required reboiler temperature. A stripping gas, which is a dry and low-pressure gas, removes the final water from the solution. Figure 6.9 shows the effect of stripping gas on TEG concentration. Proprietary processes, such as the Coldfinger® and Drizo® processes, are also available to obtain high TEG purities. Figure 6.10 shows the possible dew-point depression with the corresponding TEG concentrations in the enhancement processes.

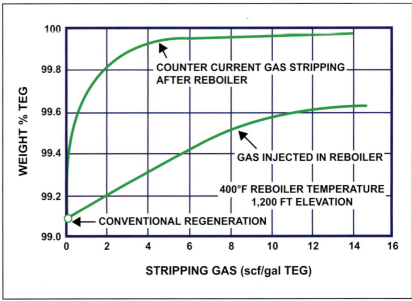

Source: GPSA Engineering Data Book, 12th edition

Figure 6.9 Effect of stripping gas on TEG concentration

Process	TEG Conc. wt%	Water Dew-Point Depression Possible, °F
Vacuum	99.2 to 99.9	100 to 150
Coldfinger®	92.2 to 99.7	100 to 130
Drizo®	99.99 to 99.999*	180 to 250*
Stripping Gas	99.2 to 99.9	100 to 150
* With Solvent Dryer		

Source: GPSA Engineering Data Book, 12th edition

Figure 6.10 Glycol regeneration processes

Plant Processing of Natural Gas

If the inlet gas cannot be dehydrated to the desired level, the following operational factors must be evaluated:

- Glycol concentration
- Glycol circulation
- Glycol solution condition

The minimum TEG concentration must be evaluated to determine that it meets the outlet gas dehydration specifications. Referring back to figure 6.1, the water dew-point temperature of the outlet gas can be determined given its maximum water content and pressure. As seen in figure 6.11, the contact temperature is about 70°F for 98% weight TEG. Using figure 6.12 and the dew-point temperature of the outlet gas previously determined and measured, the temperature of the dehydrated gas leaving the contactor and the required TEG concentration can be determined. At 1,000 psia and 0°F equilibrium water dew point, the gas contains only about 2 pounds of water per MMscf of gas.

Using figure 6.11 and the dew-point temperature of the outlet gas previously determined and measured, the temperature of the dehydrated gas leaving the contactor and the required TEG concentration can be determined. At 1,000 psia and 0°F equilibrium water dew point, the gas contains only about 2 pounds of water per MMscf of gas.

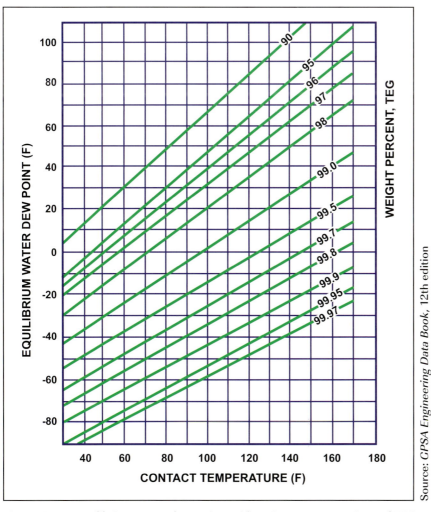

Source: GPSA Engineering Data Book, 12th edition

Figure 6.11   Equilibrium water dew points with various concentrations of TEG

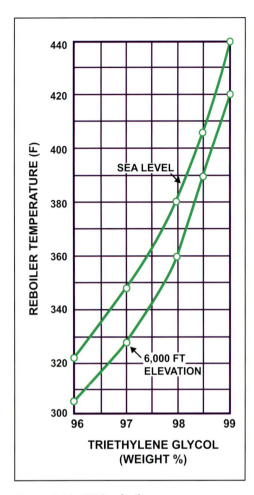

Figure 6.12   TEG reboiler temperatures

As seen in figure 6.12, the reboiler temperature will be approximately 360°F for elevation of 6,000 feet and 380°F for sea level.

Dehydration of the gas to a lower dew point than that of the hydrate formation point is required when a glycol system is upstream of a refrigeration process. The dew-point temperature achieved will be no lower than that of the refrigerant from the gas chiller. The graphs from the previous method can be used to establish the TEG concentration and reboiler temperature needed. If the reboiler temperature is higher than 400°F, the temperature at which TEG decomposes, an enhancement process must be used to get the required TEG concentration. The enhancements include vacuum, stripping gas, or some proprietary process. If a stripping gas is used, excess amounts must be avoided because this causes increased TEG losses and waste gas.

The correct level of glycol circulation is necessary for the desired level of dehydration. The lean TEG is used to determine the dew-point temperature and reboiler temperature. The rate of circulation must be sufficient to prevent excess water from rising in the gas to the top trays of the contactor and reducing the dehydration effectiveness of the lean TEG. Also, increasing the circulation rate requires less heat provided by the reboiler. The 12th edition of the *GPSA Engineering Data Book* (pages 20–34) is a good source for charts that illustrates how increasing circulation achieves higher dew-point depressions. The circulation rate rapidly rises exponentially and reaches a limit. Most economic TEG systems circulate 2–5 gal TEG/lb H_2O absorbed.

For glycol systems, there are additional problems with glycol solution conditions that must be addressed. Heavy hydrocarbons, if present in the inlet gas, are condensed by the TEG in the contactor and are difficult to separate from solution in the regenerator. This problem can be minimized if the lean TEG is 10°F–20°F warmer than the inlet gas in the contactor. Aromatic hydrocarbons, such as benzene, are very soluble in TEG. These are stripped from the solution in the regenerator and, when vented, might cause environmental or safety hazards. Because the level of absorption increases with higher pressures, lower temperatures, and high circulation rates, a taller contactor tower with decreased circulation rates minimizes this problem.

Thermal degradation of TEG affects its dehydration performance. Decomposed compounds and heavy hydrocarbons can be easily seen in the *gauge glass* of a surge tank. Heavy hydrocarbons and decomposed glycol are seen as a black layer above and below the TEG. The proper reflux rate will prevent hot spots from forming in the reboiler and resulting in the formation of degradation products.

Corrosion tends to be a major problem in glycol units. It is caused mainly by CO_2 and H_2S. The use of a pH-control chemical can minimize corrosion caused by acidic degradation products. These chemicals must be removed in the rich glycol flash. Although corrosion inhibitors can prevent corrosion, they can be an additional source of foaming.

Solid Desiccants

Solid desiccants that absorb water are characterized by their ability to dehydrate natural gas streams to low dew points. Typically, dehydration of natural gas with solid desiccants is limited to cryogenic plants where temperatures reach –100°F to –150°F.

There are three main types of materials used in drying natural gas: silica gels, alumina, and molecular sieves. In dry-bed dehydration, these materials are regenerated for reuse with hot gas followed by cooling. To maintain this continuous process, two or more towers or vessels are used. One tower is being regenerated while the others are adsorbing water.

The adsorption cycle lasts from 8 to 24 hours. The regeneration cycle, consisting of heating and cooling, follows. There are various types of desiccants used in this process.

Silica Gels and Activated Alumina

Silica gels and activated alumina are capable of achieving outlet dew points of –90°F and –60°F, respectively. Compared to molecular sieves, alumina and silica gel need less regeneration heat and a lower regeneration temperature of about 400°F. Activated alumina is a base with a high pH value. It, therefore, should not be used for natural gas streams with high acid gas concentrations. Silica gel is stable against acid damage and works well on gas streams with high acid gas concentrations.

Molecular Sieves

Molecular sieves are a popular method for dehydration because of their ability to reach the lowest dew points attainable for any adsorbent (< 0.1 ppmv). They can achieve dew points down to –300°F. Molecular sieves are commonly used in cryogenic operations, such as ahead of NGL ethane recovery plants and LNG liquefaction plants. Depending on the type of sieve used, the regeneration temperatures range from 425°F to 550°F.

Alternative methods for natural gas dehydration include calcium chloride, refrigeration, and membranes. Calcium chloride is used as a nonregenerable-type desiccant for dehydration and can achieve outlet water contents of 1 lb/MMscf. The typical calcium chloride capacity is 0.3 $CaCl_2$ per pound of water. This operation is a valuable alternative to glycol units when used at low gas rates and at remote dry wells. If properly protected against hydrate formation, natural gas can be dehydrated down to –150°F. Membranes can be used to dehydrate natural gas streams, but they are generally only considered for plants that use low pressure and low-heating value fuel. Typical solid desiccant properties are tabulated in figure 6.13.

Figure 6.13   Typical desiccant properties

Desiccant	Shape	Bulk Density, lb/ft^3	Particle Size	Heat Capacity, Btu/(lb•°F)	Approx. Minimum Moisture Content of Effluent Gas
Alumina Alcoa F200	Beads	48	7x14 Tyler Mesh ⅛" / ³⁄₁₆" / ¼"	0.24	-90°F Dew Point
Activated Alumina UOP A-201	Beads	46	3-6 Mesh or 5-8 Mesh	0.22	5-10 ppmv
Mole Sieve Grace–Davison 4A	Beads	42-45	4-8 Mesh or 8-12 Mesh	0.23	0.1 ppmv (-150°F)
Molecular Sieve UOP 4A-DG	Extrudate	40-44	⅛" or ¹⁄₁₆" pellets	0.24	0.1 ppmv
Mole Sieve Zeochem 4A	Beads	45-46	4-8 Mesh or 8-12 Mesh	0.24	0.1 ppmv
Silica Gel Sorbead® – R	Beads	49	5x8 Mesh	0.25	-60°F Dew Point
Silica Gel Sorbead® – H	Beads	45	5x8 Mesh	0.25	-60°F Dew Point
Silica Gel Sorbead® – WS	Beads	45	5x8 Mesh	0.25	-60°F Dew Point

Source: GPSA Engineering Data Book, 12th edition

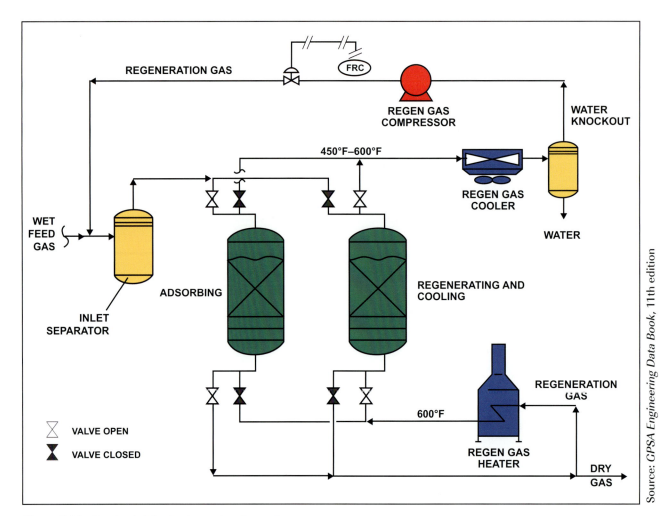

REGENERATION GAS

FRC

REGEN GAS COMPRESSOR

WATER KNOCKOUT

450°F–600°F

REGEN GAS COOLER

WET FEED GAS

INLET SEPARATOR

ADSORBING

REGENERATING AND COOLING

WATER

VALVE OPEN

VALVE CLOSED

600°F

REGENERATION GAS

REGEN GAS HEATER

DRY GAS

Source: GPSA Engineering Data Book, 11th edition

Figure 6.14   Dry-bed dehydration unit schematic

Natural gas usually flows downward through the bed in the adsorption cycle, and regeneration gas flows upward when the bed is in the regeneration cycle. Figures 6.14 and 6.15 show a simple two-tower system that can be used for each adsorbent type. In this process, after passing through the inlet separator, the wet feed stream moves down the adsorber bed until the desired dew-point specification is met and the dry gas exits the bottom.

When being regenerated, the upflow of hot gas drives off the water and some heavy hydrocarbons are adsorbed. After the water is removed, the cool gas stream passes upward through the bed to arrive at the required temperature before the adsorption cycle repeats (fig. 6.16). The regeneration gas flows through a gas cooler where the water is condensed and collected. The gas is recompressed and recombined with the wet inlet feed. This process can be easily automated for trouble-free operation.

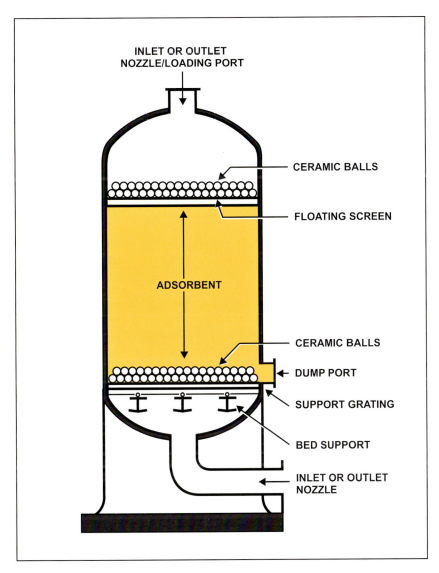

INLET OR OUTLET
NOZZLE/LOADING PORT

CERAMIC BALLS

FLOATING SCREEN

ADSORBENT

CERAMIC BALLS

DUMP PORT

SUPPORT GRATING

BED SUPPORT

INLET OR OUTLET
NOZZLE

Figure 6.15   Diagram of an adsorption tower

Source: Oilfield Processing of Petroleum, Vol I:
Natural Gas, Manning & Thompson (PennWell, 1991)

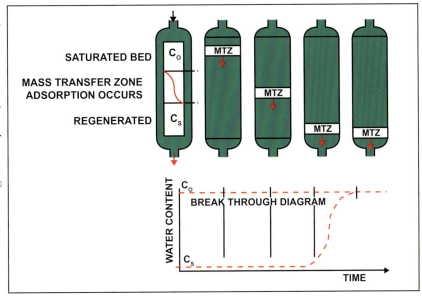

SATURATED BED

MASS TRANSFER ZONE
ADSORPTION OCCURS

REGENERATED

C_o

MTZ

C_s

MTZ

MTZ

MTZ

MTZ

WATER CONTENT

C_o

BREAK THROUGH DIAGRAM

C_s

TIME

Figure 6.16   Mass transfer zone
for water solid bed adsorption
scheme

Dehydration and Mercury Removal

Design Issues

There are several design issues that impact the dehydration processes.

Poor Outlet Water Dew Point

In an outlet stream, a bed can only produce a dew point in equilibrium with the last adsorbent particles. If sudden changes occur, the following should be examined:

- Regeneration gas temperature
- Regeneration gas flow rate
- Flow distribution/channeling
- Oxygen concentration
- Bed fouling
- Feed bypass
- Internal insulation bypass
- Hygrometer

The *thermocouples* should be examined to make sure that the readings are correct. Typically, the regeneration outlet temperature is 10% less than the inlet temperature due to heat loss from radiation and conduction. The flow meter should be calibrated and checked to see that it is working correctly. The regeneration gas must be distributed uniformly throughout the bed. To check uniform distribution of regeneration gas, the pressure drop across the bed should be above 0.01 psi/ft of bed. In general, the oxygen content should be below 20 ppmv to avoid problems from the formation of CO_2, water, and sulfur. Bed fouling can also be prevented by installing an efficient inlet separator.

The installation of a multi-stage, high-efficiency filter separator is strongly recommended to maximize desiccant life. The separator should be large enough to handle the most demanding conditions. It should have adequate capacity to handle free liquids and solids and contaminants such as compressor oils, *absorption oil,* liquid hydrocarbons, *paraffins,* corrosion inhibitors, glycol, amines, pipeline rouges, iron sulfide, iron oxide, sand particles, drilling muds, pipeline scale, and elemental sulfur.

These filter separators consist of a primary section and a final, high-efficiency separation section, which coalesces and separates the remainder of the liquid. In the primary section, entrained liquids and foreign contaminants are removed. The inlet section is designed to use small-diameter cyclones created by gravity and centrifugal force to remove liquid and solid particles. In some applications, a type of vane-type mist extractor can be used instead. By removing the bulk of the entrained and carryover liquids in the primary section, this design increases the life of the final separation elements and holds the pressure-drop buildup to a minimum.

The final separation section has one or more cylindrical coalescing elements mounted vertically or horizontally on support tubes. The gas and fine mists pass from the inside to the outside of the elements. In passing through the coalescing elements, the entrained mist particles diffuse and enter the closely spaced surfaces of the element where the mist merges into larger liquid droplets. The larger liquid droplets emerge on the outer surface of the coalescing element and run down the sides of the element to the liquid collection chamber. The gas, free of liquid particle entrainment, rises and passes out of the separator through the upper gas outlet nozzle.

Typically, these separators can remove 100% of all liquid particles above 3 microns. Depending on design conditions, they can remove up to 99.98% of all particles less than 3 microns. This efficiency is maintained throughout the entire flow capacity range, which includes the maximum design flow capacity (fig. 6.17).

Courtesy of Burgess Manning, Inc.

Figure 6.17   Horizontal filter separator

Adsorbent Life

In a well-designed plant, the adsorbent (molecular sieve) life ranges from 2 to 5 years, with 3 years being the average. The adsorbent depends on feed compositions, efficiency of the inlet separator, and the regeneration conditions. Due to the continuously thermal cycling at high and low temperature associated with each regeneration step, the number of regeneration cycles is a major factor in the useful life of the bed.

Degradation is a problem associated with the loss of capacity of the solid desiccant to absorb water. It occurs if the bed becomes contaminated by glycol, amine, heavy oils, and elemental sulfur or other contaminants. Elemental sulfur decreases the capacity of the adsorbent. It is produced by a reaction of hydrogen sulfide with the desiccant itself. Additionally, slugs and the entrainment of liquid water and hydrocarbons promote rapid degradation of the beds.

Pressure Drop

Over their useful life, molecular sieves used in natural gas dehydration show an increasing pressure drop. This is due in part to the normal accumulation of materials on the adsorbent particles that cannot be removed without damage during regeneration. If aqueous salts come into contact with the bed, they cause the adsorbent particles to burst during regeneration, resulting in increased pressure drops across the bed. Also, in poorly designed units, liquid refluxing results in water condensing at the top of the regeneration bed. Then, the water rains back onto the bed, destroying the bed when it starts to boil.

Fines, such as dust and particles, are the main cause for an increasing bed pressure drop. Fines are produced during the mechanical destruction of the molecular sieve beds from high gas velocities, gas channeling, and bed fluidization.

Switching Valves for Adsorption and Regeneration Operation

Due to temperature swings from ambient to as high as 600°F and the accumulated dust from the desiccant degradation, the quality of the switching valves is critical. The most reliable valves are ball valves with metal seats that offer tight shutoff, even at operating temperatures between 40°F and 550°F. Although less frequent, other design considerations are:

- COS formation resulting from H_2S in the inlet gas. The COS passes through to the downstream NGL recovery plant, requiring the need for NGL product treating to resolve this problem.

- Hydrocarbon retrograde condensation across the beds due to the bed pressure drop. Such condensation generally disrupts the adsorption of water by loading up the mol sieve with heavy hydrocarbons that are difficult to regenerate.

Mercury Removal Unit (MRU)

The main purpose of the *mercury removal unit (MRU)* is to remove trace quantities of mercury found in the gas feed to the aluminum heat exchangers of an NGL recovery unit. Some NGL recovery processes use aluminum heat exchangers such as the *brazed aluminum heat exchanger (BAHX)* or a core and shell heat exchanger. Mercury, even in trace quantities, will rapidly corrode aluminum. The mercury damage to aluminum heat exchangers causes corrosion, leakage, shutdown, and efficiency problems.

Some objectives for mercury removal are to:

- reduce the levels of mercury that workers are exposed to during maintenance and operation of equipment, pipelines, and process piping;

- meet environmental emission limits;

- reduce potential liabilities from mercury-contaminated product streams;

- prevent poisoning of precious metal catalysts for some end-users of natural gas.

The location of an MRU in a gas plant depends on factors such as waste disposal costs resulting from mercury contamination of equipment. To minimize contamination, MRUs should be located close to the plant inlet. However, many MRUs are placed after the dehydration unit because of the wide use of sulfur-impregnated carbon systems that require a condensate-free, dry gas feed. A process flow diagram in which mercury removal takes place after molecular sieve gas dehydration is shown in figure 6.18.

As shown in figure 6.18, the dry gas from the dehydration unit flows through non-regenerable packed beds in the mercury adsorbers. Downstream of the adsorption beds, the gas is passed through a cartridge filter to remove any entrained bed material and is then sent to the NGL recovery unit.

No bypass is provided around the mercury adsorbers. This ensures that the NGL recovery unit cannot be operated without the mercury removal unit on-line, thereby safeguarding the aluminum heat exchangers downstream.

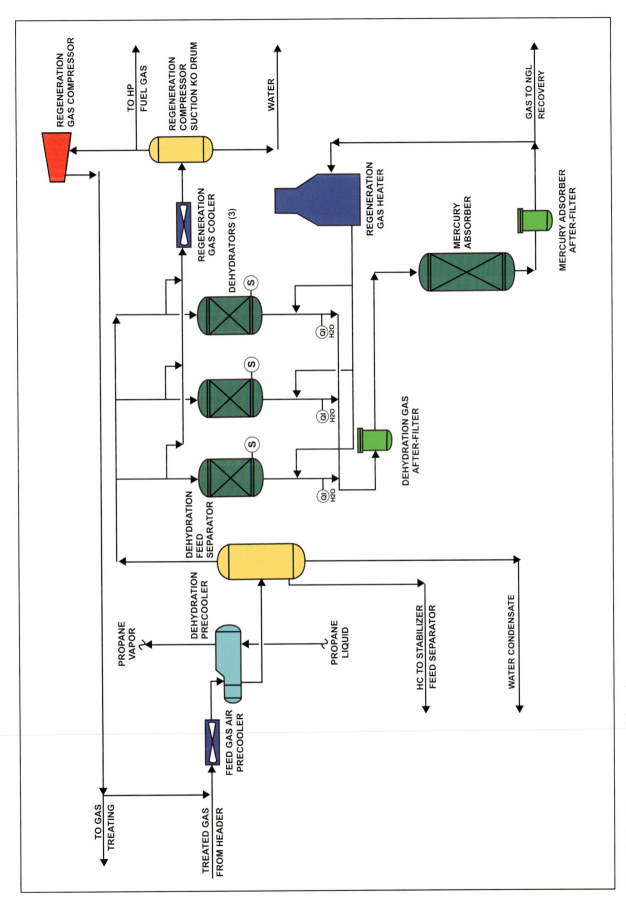

Figure 6.18 Mercury removal flow diagram

Design Basis and Specifications

Gas Dehydration Unit

The total amount and concentration of mercury in natural gas depends on the geographic location of gas fields. Low levels of mercury at about $0.01\mu g/$ Nm^3 are reported in North America. High levels of mercury at about $6000\,\mu g/$ Nm^3 are reported in Northern Germany. The raw feed from pipelines comes out of individual wells from different gas-gathering systems. In planning the design cases, engineers must communicate with pipeline and gas-production suppliers to study the feed gas flow, composition, and impurities.

The mercury content of the feed gas to the MRU is determined by its position in the process flow scheme. If the MRU is placed after the gas dehydration unit, the small amount of mercury present in the raw pipeline gas is generally removed in earlier gas-treating and drying units. In gas treating, the mercury removed will contaminate the acid gas. In drying units, the mercury removed will contaminate the regeneration gas. However, because the mercury concentration is small, it usually does not present any problems.

A specification of $0.01\ \mu g/Nm^3$ or < 0.001 ppb (1.0 ppt) in the treated gas exiting the MRU is common and appears to provide adequate protection against mercury corrosion.

General Considerations

Mercury poses a threat to plant operations because it corrodes aluminum components in the cryogenic heat exchangers. To prevent damage to equipment, mercury must be removed. The most common type of mercury corrosion is *amalgam* corrosion, which occurs when mercury and aluminum form an amalgam mixture in the presence of water. Metal embrittlement, which does not involve water and is faster than amalgam corrosion, is controlled by liquid mercury diffusion into the grain boundaries of stressed alloys. The mercury adsorption weakens the metal and results in cracks along the grain boundary.

Technology Selection

Gas-phase mercury removal systems use (1) a sulfur-impregnated activated carbon, (2) a metal sulfide on carbon or alumina, and/or (3) a regenerative molecular sieve bonded with a metal that *amalgamates* with mercury. All three systems have an inert substrate, or support, which is chemically or physically bonded to a reactive compound that forms a stable mercury compound retained by the bed.

The most commonly used system for the removal of mercury is the sulfur-impregnated activated carbon. However, the metal sulfide systems have some advantages. Regenerative molecular sieve systems have been installed to treat LPG and natural gas.

To select the most efficient system, an economic study is recommended. Important parameters for the economic study should include inlet pressure, inlet temperature, and inlet mercury concentration.

Sulfur-Impregnated Activated Carbon

In sulfur-impregnated carbon systems, elemental mercury physically adsorbs on activated carbon. It then reacts with elemental sulfur to form nonvolatile mercuric sulfide. There are a number of problems associated with this system:

- Activated carbon has a high surface area and small pore size. It is an effective adsorbent but is susceptible to capillary condensation by water and C_5+ hydrocarbons.

- Capillary condensation restricts access of mercury to the sulfur and increases the mass transfer zone. However, several years of continual operation have been reported with wet NGL gas using larger pore size material, such as Procatalyse's CMG 273™.

- Organic mercury also reacts with elemental sulfur but at a slower rate than elemental mercury. Organic mercury is usually not targeted for removal by this mercury removal system.

- Sulfur is lost by conversion from a solid to vapor and leaching by liquid hydrocarbons, which reduces capacity and leads to fouling of downstream equipment.

- Mercury might need to be removed by a thermal process if the spent adsorbent cannot be sent to a landfill.

Companies providing sulfur-impregnated carbon systems include Calgon, SME, Norit, Procatalyse, Gas Land, and Carbontech. With the exception of Procatalyse, these vendors prefer to locate separate beds downstream of the dehydration unit. Procatalyse requires a gas preheater to locate the mercury bed upstream of the dehydration beds. The majority of the gas plants use the Procatalyse system for economic reasons and concern about variable mercury inlet concentration. Even with additional vessel costs, the overall *total installation cost* (TIC) is usually lower than average.

Metal Sulfide on Carbon or Alumina

In *metal sulfide systems,* a reactive metal bonded to a support structure, such as carbon or alumina, forms a reactive metal sulfide by reaction with H_2S in either the hydrocarbon gas or liquid to be treated. The metal sulfide forms *in situ,* or in place, without having to be removed from the equipment. The metal sulfide can also be supplied preactivated. The metal sulfide reacts with mercury to form mercuric sulfide.

The advantages of metal sulfides include insolubility in liquid hydrocarbon and less sensitivity to water. This unit can be located upstream of the gas-treating unit or dehydration unit. It also can be located downstream of the dehydration unit. However, this system uses PURASPEC™, which is a costly treatment. Additional vessels are needed, but these costs are small compared to impregnated-sulfur on carbon systems used by some gas and LNG plants.

Regenerative Molecular Sieve—Silver (Ag) on Zeolite

In *regenerative molecular sieve systems,* such as UOP's HgSiv™ or Zeochem's T-2615Z, silver is deposited on the outer surface of the molecular sieve pellets. The silver then reacts with mercury from the process fluid, either gas or liquid, and forms an amalgam that is insoluble in hydrocarbon liquid. These regenerative molecular sieve adsorbents are put at the vessel bottom of the dehydrating unit. Because of limited adsorbent capacity, these systems require a regeneration system. Heating in the regeneration cycle releases the mercury. The regeneration gas containing the mercury vapor then goes to a secondary MRU.

This process needs an adequate three-phase separator to get rid of the liquid mercury after regeneration. Moreover, the water condensed might contain mercury. This system also needs an additional sulfur-impregnated activated carbon vessel to treat the mercury in the regeneration gas.

Design Considerations

MRU Location

The MRU location depends on feed composition, plant configuration, and the products being marketed. When designing and locating an MRU, it is important to understand how mercury and mercury compounds are distributed in the feed gas to the plant. Most, if not all, of the mercury in natural gas is believed to be in elemental form. Organic or inorganic mercury compounds dissolved in hydrocarbon liquids and water are probably present. Whether mercury compounds are associated with corrosion problems in the gas plants remains unresolved. Therefore, most MRUs are designed to remove the mercury in its elemental form.

Three locations for the MRU have been reported or suggested:

- Immediately after the dehydrator
- After the acid gas removal unit
- Before the acid gas removal unit

All the mercury removal systems discussed can be placed immediately after the dehydrator. However, not all systems can be placed before dehydration. If the MRU is being integrated into an existing plant, there might be size and piping limitations at some locations. Existing plant data can be used to assess the costs associated with different possible locations.

Mercury Removal Costs

The capital cost for a mercury-removal unit is small when compared to overall plant costs. However, the cost of mercury contamination of aluminum heat exchangers can be high. High production losses result from total shutdown for long periods of time while heat exchangers are replaced. Economically, it is advisable to include a mercury-removal unit even if gas analyses indicate that no mercury is present.

Free-Liquid Removal

Free-liquid removal is critical for all sulfur-impregnated carbon systems used in mercury removal. All existing plants with a mercury-removal system have free-liquids separators. Additionally, some plants have installed an extra coalescing filter between the free-liquids separator and the mercury-removal beds to remove the entrained water particles. Coalescing filters have extremely fine mesh or holes and have the capability of removing oil from the stream.

Pressure Drop

The pressure drop of a mercury-removal system should be considered in an economic evaluation, but it is not a major issue. The metal sulfide system with PURASPEC™ has the lowest pressure drop compared to the sulfur-impregnated carbon and regenerated molecular sieve systems.

Superficial Velocity and Residence Time

The superficial velocity and the residence time should be within the acceptable range for any mercury-removal unit. Removal units that can process higher allowable superficial velocity and lower allowable residence time are preferred because of savings on the initial total installation cost. However, if vendors cannot meet their claims, then either the bed life will be shorter or the absorbent cannot meet the mercury outlet specification.

Equipment Selection and Design

Mercury Adsorber

An MRU consists of adsorbent beds that remove the mercury by direct chemical reaction. It is difficult to design or even accurately predict the expected bed life. The critical parameters to determine the bed size and efficiency are:

- volume of the mass transfer zone (MTZ);
- rate of movement of the MTZ;
- dynamic capacity (exhaustion rate).

Spent adsorbent must be treated as a hazardous waste. It can be sold to metal refiners or adsorbent suppliers who can recover the mercury. If it is not sold to a processor, it must go to an approved landfill designated for the disposal of mercury components.

Mercury Removal After-Filter

The function of the mercury adsorber after-filter is to remove any carryover MRU dust, particles, and fines. The removal protects the heat exchangers downstream from fouling caused by carryover MRU fines. The filters are offered by many vendors.

Case Study

Incorrect Coalescer Liquid Level

An MRU with a nonregenerable bed of Procatalyse's CMG 273 was installed before acid gas removal to remove mercury from natural gas. After a few hours of operation, the performance degraded. The bed material was analyzed and met quality control expectations.

Later, heavy natural gas condensate was found upstream and downstream of the MRU. It turned out that the feed gas was short circuiting the second stage of the coalescer separator and pushing slugs of hydrocarbon liquid from the unit into the MRU. Resetting the correct liquid level in the coalescer-separator stopped the feed gas from pushing liquid into the MRU, and the dried gas freed the CMG 273 material from the condensate.

REFERENCES

Cameron, C.J., Bartel, Y., and Sarrazin, P., "Mercury Removal from Wet Gas," 73rd Annual GPA Convention, New Orleans, Louisiana, March 7–9 (1994).

Campbell, J.M., "Absorption Dehydration and Sweetening," Gas Conditioning and Processing, Campbell Petroleum Services, Norman, Oklahoma (1976).

Carnell, P.J.H., Row, V.A., and McKenna, R., "A Re-think of the Mercury Removal Problem for LNG Plants," 15th International Conference & Exhibition of Liquefied Natural Gas, Barcelona, Spain, April 24–27 (2007).

The Dow Chemical Company, *Gas Conditioning Fact Book*, Midland, Michigan (1962).

GPSA Engineering Data Book, 12th Edition, Gas Processors Suppliers Association, Tulsa, Oklahoma (2004).

Huffmaster, M.A., *Gas Dehydration Fundamentals*, Laurance Reid Gas Conditioning Conference (LRGCC), Norman, Oklahoma (2004).

Kohl, Arthur L., and Nielsen, R., *Gas Purification*, 5th edition, Gulf Professional Publishing, Elsevier, Inc. Burlington, Massachusetts (1997).

Malino, H.M., "Fundamentals of Adsorptive Dehydration," Laurance Reid Gas Conditioning Conference (LRGCC), Norman, Oklahoma (2004).

Manning, F. S., and R. E. Thompson, *Oilfield Processing of Petroleum*, Vol. 1, Natural Gas, PennWell Publishing Company, Tulsa, Oklahoma. (1991).

Smith, R.S., "Fundamentals of Gas Dehydration Inhibition/Absorption Section," Laurance Reid Gas Conditioning Conference (LRGCC), Norman, Oklahoma (2004).

Spiric, Z., "Innovative Approach to the Mercury Control during Natural Gas Processing," Engineering Technology Conference on Energy, Houston, Texas, February 5–7 (2001).

Trent, R. E., "Dehydration with Molecular Sieves," Laurance Reid Gas Conditioning Conference (LRGCC), Norman, Oklahoma (2004).

Wilhelm, S.M., McArthur, A., and Kane, R.D., "Methods to Combat Liquid Metal Embrittlement in Cryogenic Aluminum Heat Exchangers," 73rd Annual GPA Convention, New Orleans, Louisiana, March 7–9 (1994).

Wilhelm, S.M., "Design Mercury Removal Systems for Liquid Hydrocarbons," Hydrocarbon Processing (April 1999).

Straight from the well, natural gas is a mix of hydrocarbons, including methane, ethane, propane, and butane. It also contains many nonhydrocarbons, such as nitrogen, helium, carbon dioxide, hydrogen sulfide, and water. Raw gas is processed to separate these components. These processes continue to improve with advanced technology.

Previously, when kerosene was a highly valued product and natural gas was simply an unwanted byproduct, most natural gas was wasted. Gas was commonly flared off—a process that often continued day and night. Little was done to capture any gas products. A few operators put traps in their lines to catch *drip gas*. A form of gasoline, drip gas naturally condenses as it comes from the natural gas wells and cools in field-gathering lines.

Gasoline began to surpass kerosene sales around 1912. The first gas processors learned to increase the yield of drip gas by compressing the gas and allowing it to cool. With the demand for fuel gasoline growing quickly, producers began trying to get more usable products from the oil and gas. An increased-yield process, called *lean oil absorption,* was developed in the 1920s. Lean oil is a hydrocarbon liquid, usually lighter than kerosene. In contact with natural gas, lean oil absorbs some of the heavier hydrocarbons from the gas, which can be separated from the oil later. Using this process, operators recovered more gas condensate as well as butane, a gas that liquefies easily under pressure. This was the beginning of the NGL market.

In the 1950s, processors improved the yield from lean-oil absorption by adding refrigeration to the process. Advancing technology has added the development of better refrigeration equipment, lower processing temperatures, and new ways to market natural gas.

Lean-oil absorption is only one of many ways to separate the various products in natural gas. Instead of using lean-oil absorption, plants might use a less expensive process by refrigerating the gas to remove the propane and heavier hydrocarbons. Many of the newest gas processing plants only produce a single NGL product called Y-grade, which is then shipped to another plant for further separation.

Chapter 7 presents the older, but still commonly used, lean-oil absorption process. Chapter 8 covers the newer processes of refrigeration and *turboexpansion*. However, the complete range of process possibilities cannot be fully explored in these chapters because gas plants often use combinations and modified versions of these processes, depending on the range of products produced.

7
NGL Recovery— Lean-Oil Absorption

LEAN-OIL ABSORPTION

The three primary systems in an oil absorption plant are *recovery*, *rejection*, and separation (fig. 7.1). There are many ways to configure a gas processing plant, depending on the nature of the gas being processed and market demands. In this section on the lean-oil absorption process, examples are used based on a typical refrigerated-absorption plant designed to recover ethane. Other plants using lean-oil absorption can be designed to produce more propane, butane, pentane, or natural gasoline.

In the recovery phase, products are removed from the inlet gas by a combination of refrigeration and oil absorption. The process, however, is not exact. After the first pass, some unwanted methane always remains in the recovered product and must be removed before it goes to the customer. The rejection processing system eliminates the remaining methane from the product stream. This leaves only the desired product and absorption oil to handle in the third processing system, separation.

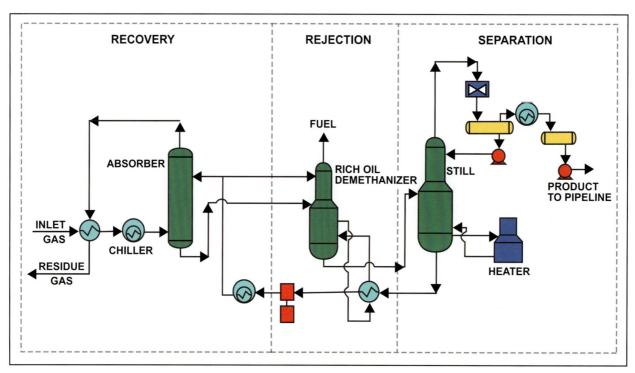

Figure 7.1   Oil absorption plant systems

Source: *GPSA Engineering Data Book*, 12th edition

The Recovery System

Recovery is the first of three processing steps in a lean-oil absorption plant. In the recovery system, all product to be recovered is removed from the inlet gas. The main piece of equipment is the *absorber*. Most of the remaining equipment in the recovery system is used to increase the operating efficiency of the absorber.

Absorption

Absorption is a process in which a liquid solvent removes some of the components from a gas. Absorption oil is called lean oil. It contains little of the type of desired hydrocarbon to be extracted from the gas.

The physical properties of lean oil vary with each plant. Many plants use Varsol®, a liquid solvent that is heavier than kerosene and lighter than paint thinner. Lean oil enters the top of the absorber and cascades down over a series of trays through a steady flow of gas. By the time oil reaches the bottom of the absorber, the light oil has become *rich oil*. It is termed rich oil because it is plentiful with hydrocarbons recovered from the gas.

In an absorption tower, gas and oil compete to hold hydrocarbons. An absorbed hydrocarbon behaves like a liquid, even though it was taken from the gas stream passing through the absorber. Hydrocarbons with lower vapor pressures, such as gasoline and butanes, are more easily absorbed than hydrocarbons with higher vapor pressures, such as propane, ethane, and methane. Hydrocarbons with higher vapor pressures are more difficult to absorb because they tend to remain in the gas phase longer. The efficiency of an absorption tower is influenced by pressure, temperature, quality of the lean oil, and the ratio of lean oil to gas.

High pressures make absorption more efficient than at low pressures. This is because the vapor pressure of rich oil under pressure is almost the same as the vapor pressure of the hydrocarbons. The pressure similarity makes it easier for the hydrocarbons to leave the gas and join the absorption oil as a liquid.

Absorption is more efficient at low temperatures because the vapor pressure of a hydrocarbon increases with heat. To lower the vapor pressure of a hydrocarbon, its temperature must be lowered. The lower the vapor pressure, the easier it is for hydrocarbons to leave the gas and become a part of the absorption oil. Conventional plants cool their lean oil and inlet gas to between 0°F and –40°F, and most use propane as the refrigerant. Inlet gas is first cooled in the gas/gas exchanger by the residue gas leaving the absorber.

The cooler inlet gas is then refrigerated to the desired temperature in the gas chiller. Inside the chiller, gas flows through the tubes while refrigerant surrounds the tubes inside the chiller shell. As cold inlet gas enters the bottom of the absorber, chilled lean oil is fed to the top.

Why Absorbers Work

Absorption is the transfer of hydrocarbon molecules from the gas phase in the inlet gas to the liquid phase or rich oil. The amount of NGL that can be absorbed at a certain pressure and temperature depends on the number of lean-oil molecules.

Lean oil has a low molecular weight. The lower the molecular weight, the lighter will be the oil. A problem is that light oil vaporizes more easily into the residue gas. Gas operators must find the lean oil best suited for the properties of the natural gas being processed and the type of equipment in the plants. The lean oil chosen should have the most molecules available for absorption and should also have characteristics that minimize the loss to vaporization.

When operators determine the correct lean-oil molecular weight for a specific plant, an oil-to-gas ratio measures how hard the absorber is working. The oil-to-gas ratio is usually expressed as the *gallons per minute (gpm)* of lean oil circulated per million cubic feet of gas through the absorber. The gas volume measured at 60°F and 14.7 psia is one *million standard cubic feet per day (MMscf/d)*.

The effect of the lean-oil-to-inlet-gas ratio is shown in figure 7.2. Note that the percentage yield of propane and ethane goes up when the amount of lean oil is increased for every million standard cubic feet per day of gas. This is because at higher ratios, more lean-oil molecules are available to absorb gas molecules.

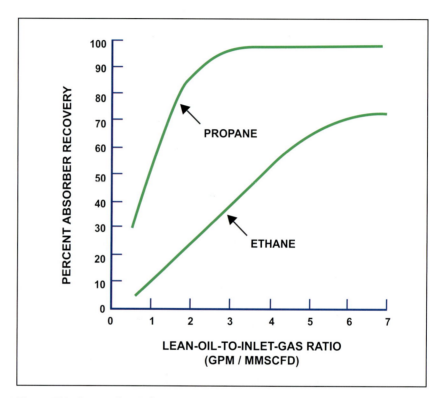

Figure 7.2 Lean-oil-to-inlet-gas ratio

Presaturation

As hydrocarbon products such as methane, ethane, and propane make the phase change from gas to liquid and become part of the oil inside the absorber, they give up their latent heat. This increases the temperature of both the oil and gas. As a result, these higher temperatures lower the efficiency of the absorber.

There are methods to remove some of the heat of absorption and minimize the increased temperature inside the absorber. One common method is a *presaturation* system (fig. 7.3). It is an effective way to hold down temperatures in the absorber. One drawback is that more lean oil must be circulated to recover the same amount of product.

In the presaturation system, the low-pressure, methane-rich vapors from the rich-oil methanizer are mixed with the lean oil and chilled then separated from the presaturated lean oil in an accumulator (fig. 7.4).

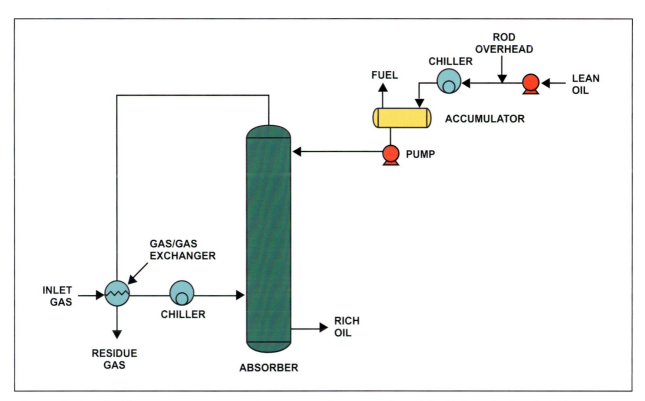

Figure 7.3   A typical low-pressure presaturation system using vapors from rich-oil demethanizer (ROD)

Courtesy of Uraltechnostroy Corporation, LLC.

Figure 7.4   Accumulator

NGL Recovery—Lean-Oil Absorption

Potential Problems

Problems in the recovery system are often due to flooding or freezing in the absorber. When an absorber floods, some or all of the lean oil fed to the absorber will leave the top of the absorber with the residue gas. As shown in the flow diagram in figure 7.5, any carryover from the lean-oil absorber will be caught by the residue gas scrubber. A *high-liquid-level alarm (HLA)* installed on the scrubber alerts the operator when lean oil is being carried over and threatening to flood the absorber.

Many plants have a *differential pressure indicator (DPI)* to measure the pressure drop across the absorber. When absorbers flood, there is a large increase in the differential pressure. The DPI detects absorber flooding before it triggers the high-level alarm on the residue gas scrubber.

Three things can cause an absorber to flood:

- A sudden, large increase in either gas or lean-oil flow
- Too much lean oil for the volume of gas being handled
- Dirt or freezing that plugs the gas or lean-oil flow paths

Sometimes the absorber must handle a sudden, large increase in gas volume. That can happen whenever a large inlet gas compressor is put on the line. If the absorber is already near its capacity, the lean-oil flow to the absorber should be reduced at that time. After the absorber has had time to settle from the increase in gas volume, the flow of lean oil can gradually be increased.

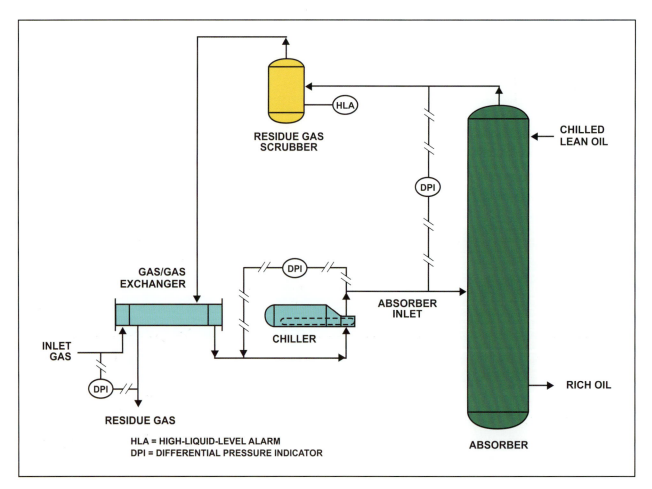

Figure 7.5   A residue gas scrubber diagram

Plant Processing of Natural Gas

The inlet gas and lean-oil capacities of an absorber are interdependent. Assume that an absorber has a lean-oil flow capacity of 1,000 gallons per minute when the inlet gas volume is 200 million standard cubic feet per day. At 250 MMscf/d, flooding will occur if the lean-oil flow is greater than 800 gpm.

When an absorber floods, the operators must greatly reduce the lean-oil flow to the absorber. The reduction of lean-oil flow to the absorber quickly stops the flooding. The operators can then determine the cause of the flooding.

When absorbers freeze, part of the normal flow space is occupied by hydrates, an icy mix of hydrocarbons and water. The blockage caused by hydrates increases the pressure drop across the equipment. Figure 7.6 plots several differential pressure indicators that an operator can use to quickly detect the freezing problem and its location.

Freezing in the inlet gas train also increases the gas temperature to the absorber. Injecting methanol for a temporary period can prevent additional freezing and break up the hydrates that have already formed. However, the only solution to a bad freeze is to allow the frozen area to warm up.

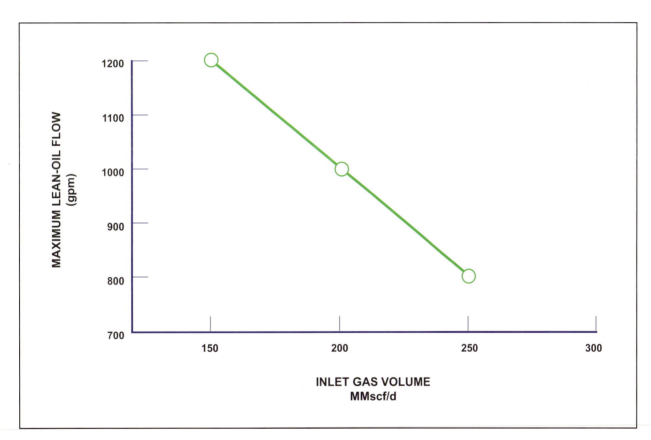

Figure 7.6   Differential pressure indicators

THE REJECTION SYSTEM

A rejection system gets rid of the methane retained by the recovery system and cannot be sold as a liquid product. Operators call it *demethanizing* the product. Usually, more than 50% of all the hydrocarbons removed from the inlet gas are methane.

To reject methane from the rich oil, the pressure is reduced and temperature is increased. This is the opposite of the high pressure and low temperature needed for adsorption. The process is a combination of rejection and reabsorbing methane. While methane is being rejected, it is important to keep the products that have been absorbed in the rich oil.

Most absorption plants reject methane in two steps. The primary methane rejection equipment eliminates at least half the methane from the rich oil. That methane is rejected by reducing pressure, increasing temperature, or both. The rejected methane usually has enough ethane and propane to make reabsorption desirable.

Hot Rich-Oil Flash Tank

There are many combinations of rejection and reabsorption. Figure 7.7 shows a hot rich-oil flash tank being used for methane rejection. Methane is rejected by heating the rich oil upstream of the flash tank. The rich oil from the absorber is pumped in at higher than inlet gas pressure, heated, and then flashed in the rich-oil flash tank. The higher pressure in the flash tank allows the flashed vapor to be recycled to the inlet gas absorber without being recompressed. The inlet gas absorber also serves as a reabsorber for ethane and heavier components in the vapor from the rich-oil flash tank.

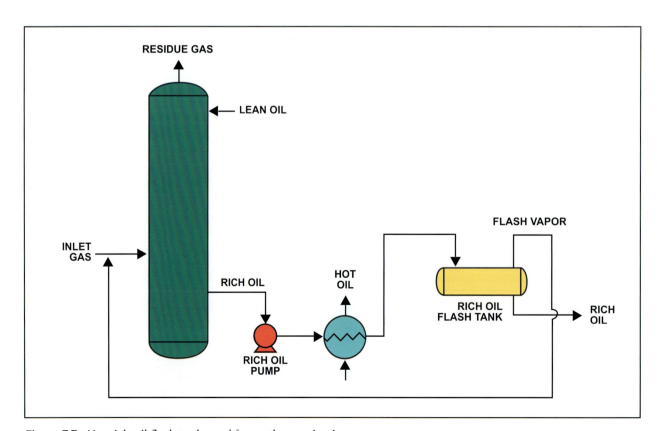

Figure 7.7   Hot rich-oil flash tank used for methane rejection

Rich-Oil Demethanizer

The second step—and heart of the rejection system—is the *rich-oil demethanizer (ROD)*. In this vessel, rejected methane leaves the top of the rich-oil demethanizer.

A rich-oil demethanizer allows rejection of the rich-oil methane, while retaining most of the ethane plus components. A flow diagram of a rich-oil demethanizer is shown in figure 7.8. The feed might come from a rich-oil flash tank, a partial demethanizer, a reboiled absorber, or straight from the main absorber. Heat is added either to the total feed or to most of the feed, while the remainder, or colder rich oil, is fed on a higher stage.

The tower performs two functions. The bottom section strips, or eliminates, by heat and vapor flow, the methane from the rich oil. The top section reabsorbs ethane and other products from vapor leaving the column.

Lean oil in the top section functions the same as in an absorber. A reboiler is used at the bottom to supply the heat that generates the vapor necessary for good stripping. The top section lean-oil feed controls the amount of ethane lost in the overhead vapor. The temperature of the rich oil leaving the stripping section, called the *rich-oil demethanizer bottoms,* controls the methane content of the rich oil.

Changes in plant operating conditions might require adjustments to the rich-oil demethanizer's bottom temperature. The pressure should be held constant. The vapor pressure of the rich-oil demethanizer bottom temperature is equal to the tower's operating pressure, so the temperature of the bottoms controls its composition. The proper bottoms composition is usually expressed as the ratio of mole percent methane to mole percent ethane.

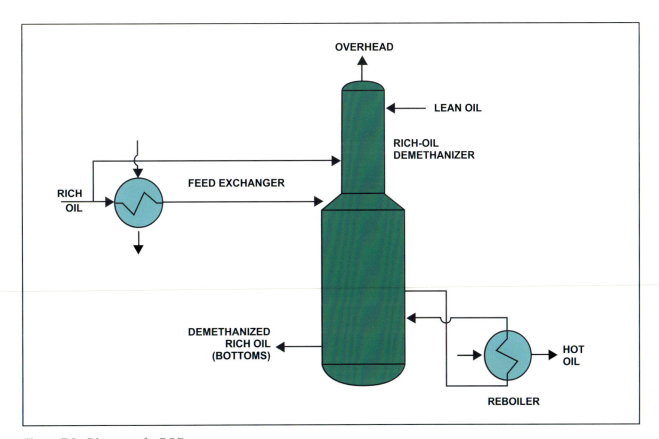

Figure 7.8 Diagram of a ROD

Increasing the bottom temperature *decreases* the methane-to-ethane ratio. Decreasing the bottom temperature *increases* the methane-to-ethane ratio. The bottoms are two components: product and lean oil. Product is lighter and has a higher vapor pressure than lean oil. It takes more heat to raise lean-oil vapor pressure than to raise product vapor pressure to the tower's operating pressure. If the bottom temperature is too high, too much product is driven overhead. If the bottom temperature is too low, too much methane is left in the bottoms.

The bottom-temperature adjustments can be compared to lean-oil-to-product ratio in the bottoms. A high ratio means a higher bottom temperature because it takes more heat to raise the lean-oil vapor pressure. A low ratio means a lower bottom temperature. Figure 7.9 is a graph of demethanizer bottom temperature versus lean-oil-to-product ratio. This graph was drawn from actual plant performance tests. While it is not accurate enough for general plant operations, the graph does show the relationship of bottom temperature and the lean-oil-to-product ratio.

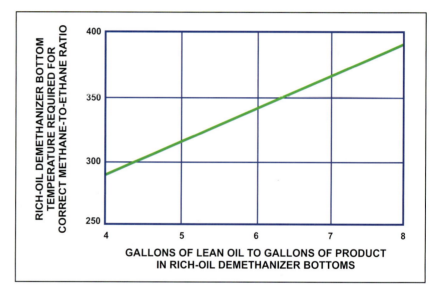

Figure 7.9 Graph of demethanizer bottom temperature versus lean-oil-to-product ratio

Plant conditions that change the lean-oil-to-product ratio require corresponding changes to the rich-oil demethanizer bottom temperature. Figure 7.10 shows some typical bottom temperature adjustments. With each adjustment, something happens to the lean-oil-to-product ratio. A *decrease* in inlet gas volume causes the plant to recover less product. Also, there will be an *increase* in the ratio of lean-oil-to-product in the rich-oil demethanizer bottoms. This adjustment requires an *increase* in the rich-oil demethanizer bottom temperature to reject the proper amount of methane from the rich oil.

REASONS FOR MAKING ADJUSTMENT	
1. Maintain methane content specification in product	
2. Recover maximum amount of ethane	

OPERATING CONDITION CHANGE	ADJUSTMENT
Decrease in inlet gas rate	Increase
Increase in inlet gas rate	Decrease
Decrease in lean-oil rate	Decrease
Increase in lean-oil rate	Increase
Decrease in processing temperature	Decrease
Increase in processing temperature	Increase
Refrigeration compressor down	Increase

Figure 7.10 Typical bottom temperature adjustments

Possible Problems

The rejected methane from the ROD can be used to presaturate lean oil, consumed as plant fuel, or recompressed to join the inlet gas or absorber residue gas. Most plants use a combination of these methods. However, all systems have a maximum volume capacity of rejected methane. When this capacity is reached, the amount of methane to the ROD must be reduced. Increasing the rejection of methane upstream of the ROD, or reducing the lean-oil flow to the absorber, will decrease the methane recovery.

Improper demethanization of the product will result in off-specification products. It will also often cause excess pressures in the product-condensing facilities and in the *still,* sometimes called the *rich-oil fractionator.* Proper demethanization depends mainly on the bottom temperature of the ROD. In many plants, heat for the ROD reboiler is hot lean oil from the bottom of the still (fig. 7.11). When that source of heat is not enough, the amount of methane to the ROD must be reduced.

Figure 7.11 The ROD reboiler is heated with hot lean oil in the bottom of the still.

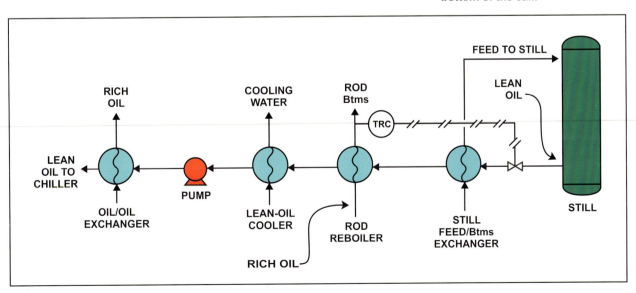

THE SEPARATION SYSTEM

Fluid in the ROD bottoms is a mix of product and lean oil. This stream is fed to the separation system, the third section of the typical plant, where the product is separated from the lean oil. The principal piece of equipment in the separation system is the still. Product goes out the top of the still and lean oil leaves the bottom. This method of separation is called *distillation*.

Distillation, the reverse of absorption, works best at high temperatures and low pressure. An efficient still allows hydrocarbons to easily leave the liquid-rich oil and become a vapor. Removing hydrocarbons from the rich oil requires a vapor to act as a stripping medium. Steam is frequently used, and stills that use steam are sometimes called *wet stills*. *Dry stills* are used to keep the lean oil free of water. Dry lean oil is used in refrigerated plants to avoid freezing or hydrate problems when the oil is chilled.

The Still

Figure 7.12 shows the simple flow diagram of a dry still. The necessary vapors are made by partially vaporizing hot oil pumped through the still's reboiler. These hydrocarbon vapors serve as the stripping medium to separate products from the lean oil.

In this system, some of the product is recirculated back to the top of the still. The technique is called reflux; it prevents lean oil from leaving the top of the still with the product.

Figure 7.12   Diagram of a dry still

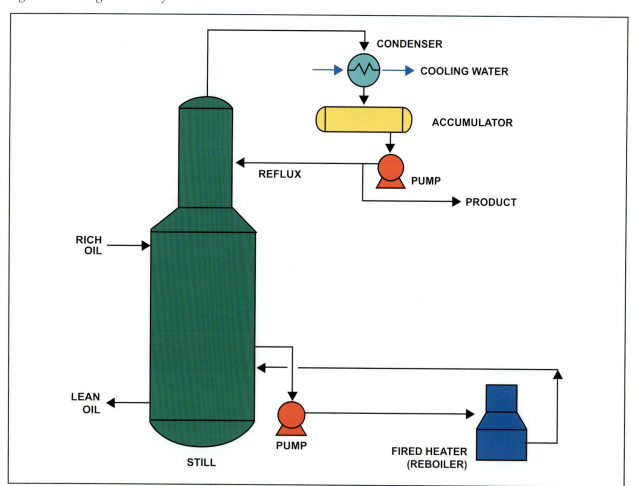

Plant Processing of Natural Gas

In figure 7.12, a fired heater is used as the still's reboiler. Plants with light lean oils, which have low molecular weight, replace the fired reboiler with either high-pressure steam or another heating medium in a shell-and-tube exchanger. Stills are normally controlled by using a constant volume of reflux and by supplying enough heat to the reboiler to maintain a fixed lean-oil outlet temperature.

The performance of a still can be and should be checked by running *American Society for Testing and Materials (ASTM International)* distillation tests on the lean oil and the product. The product sample should weather for awhile at atmospheric temperature and pressure before running the ASTM distillation test.

An operator's primary interest is the final boiling-point temperature of the product sample. This is not the *dry-point* temperature observed at the end of an ASTM distillation test. Rather, it is the observed temperature during an ASTM test when the temperature levels off as the last of the test liquid boils off. The dry point is the observed temperature as the final residue of the test dries out. A complete ASTM distillation test should be run periodically on the lean oil to measure its quality.

Good lean-oil quality is essential to the efficiency of the plant. The principal standards of lean-oil quality are the 50% boiling-point temperature and the distillation range of temperatures. When these two are correct, the molecular weight of the lean oil is fixed. At most plants, the quality of the lean-oil depends on the efficient operation of both the still and the oil purifier.

Oil Purification

A sometimes neglected part of an absorption plant is the oil purification, or oil reclaimer system. Oil reclaiming removes the heavy components that accumulate in the oil over time. Heavy components in the plant's lean oil will reduce the amount of products being recovered, increase operating difficulties, and increase lean-oil losses.

There are several types of oil-reclaiming systems. Some are batch operated, and others run automatically in a continuous feed. Figure 7.13 shows a typical design. Although operations vary from plant to plant, the reason for having oil reclaimers is the same. The purpose is to control the final boiling-point temperature of the lean oil, thereby preventing heavy ends from leaving the oil reclaimer still.

Figure 7.13 An oil-reclaiming system design

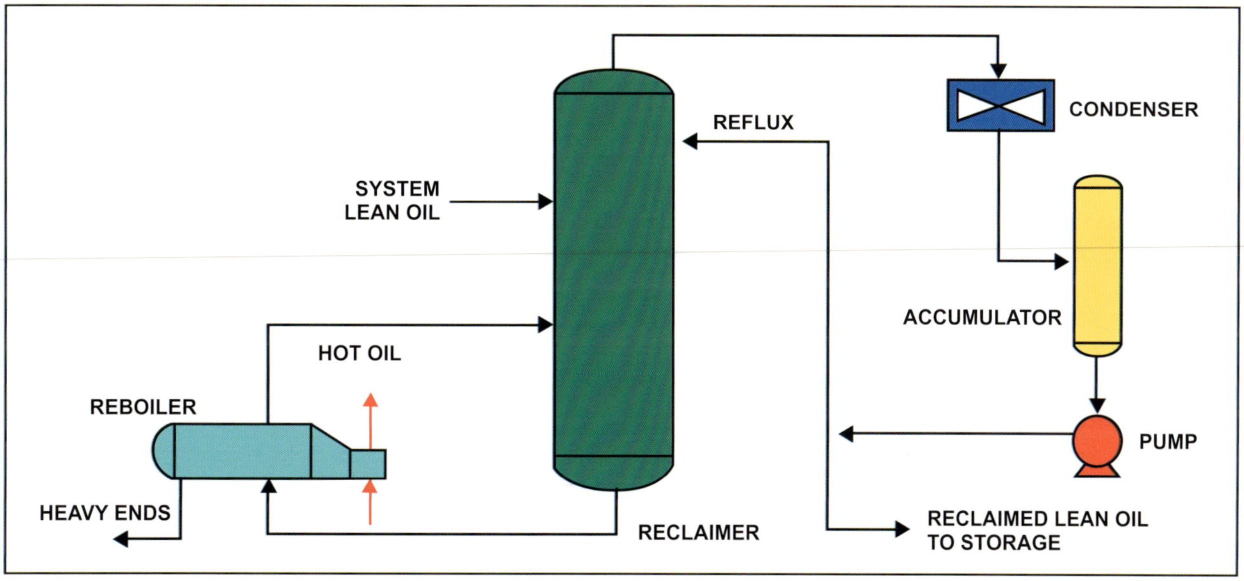

Figure 7.14 gives an example of the need to control the final boiling-point temperature of lean oil. In this example processing plant, the bad lean oil has a low initial boiling point and high final boiling point.

Because of the wide range of distillation temperatures, operations at this plant suffered until the heavy ends were removed from the lean oil by reducing the overhead temperature of the oil reclaimer. The difference in the curve shows lean oil that was within the desirable distillation range.

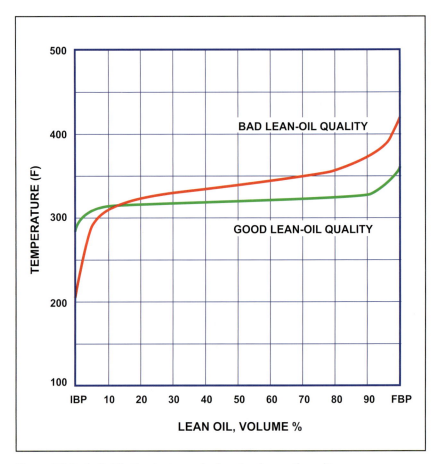

Figure 7.14   A distillation test graph showing lean-oil quality

Possible Problems

Plant operators must avoid leaving product hydrocarbons in the bottom of the still with the lean oil. If this happens, products will vaporize and leave the absorber with the residue gas instead of being contained in the lean oil. The initial boiling point of the lean oil is an effective method to measure its ability to separate products from the feed gas.

Figure 7.15 shows the losses experienced by one plant due to the poor quality of the lean oil. At a lean-oil initial boiling-point temperature of 75°F less than design case conditions, the loss of products amounted to 4.4 gallons per million standard cubic feet (MMscf) of absorber residue gas.

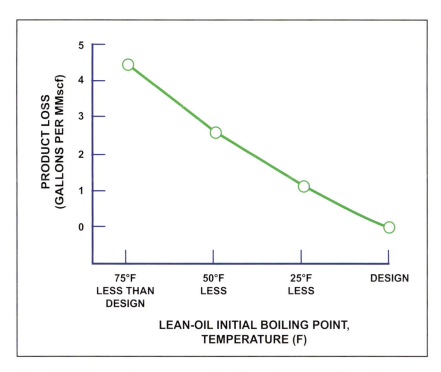

Figure 7.15   Losses due to poor quality of lean-oil initial boiling point

The final boiling point of the product is a measure of how much lean oil is leaving the top of the still with the product. Lean oil in the product can be minimized by pumping enough reflux to the still. When there is enough reflux, the final boiling point of the product should be no higher than the initial boiling point of the lean oil.

If a multipass fired heater is used for the still reboiler, it is essential that the oil flow and outlet temperatures for each pass are balanced. If either the flow or outlet temperature becomes greatly out of balance, some of the heater tubes might become dry and liquid will not flow through them. When that happens, a crusty residue of burned hydrocarbons or coke quickly forms inside the tubes and creates hot spots, which eventually melt through and cause the tubes to fail. Because replacing heater tubes is expensive and time consuming, maintaining a high circulation rate to a fired heater is critical. A high circulation rate will ensure that enough oil will remain liquid in the heater because the amount of oil vaporized is relatively constant.

REFERENCES

GPSA Engineering Data Book, 12th edition, Gas Processors Suppliers Association, Tulsa, Oklahoma (2004).

The heavier hydrocarbons, generally referred to as NGLs, might need to be removed to control the hydrocarbon dew point and/or the gas heating value. However, heavier hydrocarbons can also be a source of income for the gas producer. The history of the evolution of liquids recovery facilities from simple oil absorption to cryogenic expander processes is complex (Elliot, 1996).

Lean-oil absorption processes, such as the ambient and refrigerated processes, were commonly used until the early 1970s, as shown in figure 8.1.

8 NGL Recovery- Cryogenic

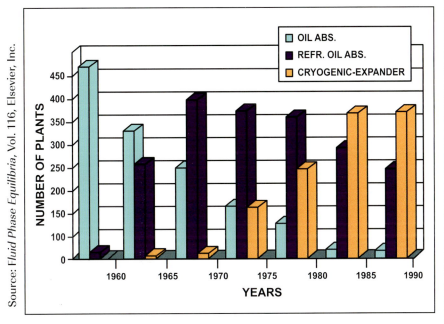

Figure 8.1   *Lean-oil absorption process and cryogenic process*

The refrigerated oil absorption process was introduced in 1957 by modifying the ambient oil absorption process to operate at lower temperatures. This allowed the use of lower molecular weight oils which recover more NGLs than the higher molecular weight oils, used in the ambient process. At a temperature of –40°F, the refrigerated oil absorption process could be used to recover up to 40+% ethane and 90+% propane in the feed gas.

A major shift in the gas processing industry began when the first low-temperature expander plant was built and brought into operation in 1964. The basic design of this plant remains today. During the 1970s, the expander plant became the dominant process scheme used to recover NGLs from feed gas. It is an efficient process for high ethane and propane recovery.

By the end of 1990, expander plants accounted for 42.5% of gas processed in the United States. During the same time, the percentage of gas processed in refrigerated oil absorption plants dropped to 32%, while ambient oil absorption plants accounted for only 3.7% of the gas processed.

Cryogenic processing requires the proper combination of pressure and low temperature to achieve the desired product recovery. Figure 8.2 shows the proper combination of pressure and temperature for 60% ethane recovery from a typical natural gas plant in the U.S. Gulf Coast. Processing at these low temperatures requires special plant techniques.

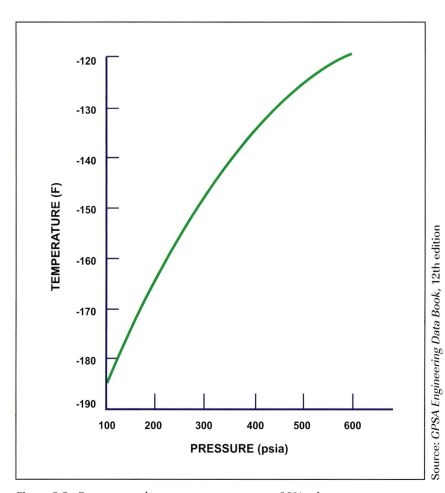

Source: *GPSA Engineering Data Book*, 12th edition

Figure 8.2   Pressure and temperature to recover 60% ethane

TYPICAL APPLICATIONS

Turboexpander Process

Turboexpanders are most widely used for the recovery of ethane and propane from natural gas. They are also used in the recovery of power from the expanding streams during the processing of helium, carbon dioxide, and hydrogen. Turboexpanders are also used in the liquefaction of helium, hydrogen, and natural gas.

The cryogenic process requires the turboexpander driving a compressor to generate low-temperature refrigeration.

The inlet gas is first prepared for processing by removing water, carbon dioxide, and other contaminants that might cause problems. Cooling of inlet gas is accomplished by using cold residue gas and cold demethanizer internal liquid streams. In cases where the gas is rich in NGLs, external refrigeration is used to supplement the cooling step. The inlet gas is divided so that part flows through the first gas/gas exchanger and is cooled by the residue gas. The remainder of the inlet gas is cooled in the demethanizer side exchanger by withdrawing cold liquid from the demethanizer. The demethanizer side exchanger serves two purposes:

- It assists in cooling the inlet gas
- It adds part of the heat necessary for properly demethanizing the plant product

The inlet gas from the first gas/gas exchanger and that from the demethanizer side exchanger join to flow through the second gas/gas exchanger. Liquid condensed from the inlet gas is separated from the vapor in the expander inlet separator. The liquid is fed to the demethanizer. The vapor flows through the expander, where the pressure is reduced, and then to the top of the demethanizer. The enlarged top of the demethanizer serves as the expander outlet separator.

Cold residue gas from the top of the demethanizer is first used to cool the inlet gas. The residue gas is then compressed to sales gas pressure by the expander-compressor and the recompressor. The expander-compressor is driven by the expander, while the recompressor driver might be a gas engine, gas turbine, steam turbine, or electric motor. Hot residue gas from the discharge of the recompressor is used for heat in the demethanizer reboiler. After final cooling, the residue gas is delivered to sales.

The solid line is the dew-point line and respresents the plant inlet gas in figure 8.3. At a fixed pressure, and if the temperature of the gas is to the right of this dew-point line, then the gas is 100% vapor. If the gas is cooled, liquid starts to condense when the temperature reaches the dew-point line. As cooling continues, more liquid is condensed until the *bubble-point* line is reached—the solid line on the left. At this point, all of the gas is liquid. Additional cooling will result in a colder liquid.

Downstream of the gas treating facilities, the inlet gas is represented by point 1 on both figures 8.3 and 8.4. Referring to figure 8.4, note that as the gas is cooled by the gas/gas exchangers and demethanizer side exchanger, its temperature moves along the dotted line to point 2. At point 2, the gas enters the expander inlet separator where the condensed liquid is separated from the vapor. This vapor now has its own pressure-temperature diagram, as represented by the dashed curve. At the expander inlet, the gas is on its dew-point line.

As the gas flows through the expander, its pressure-temperature path is shown by the dashed line from point 2 to point 3. Point 3 represents the outlet of the expander. The importance of using the expander as a driver for a compressor can be seen in figure 8.4. If the gas had just been expanded without doing any driver work, the expansion path would be from point 2 to point 4.

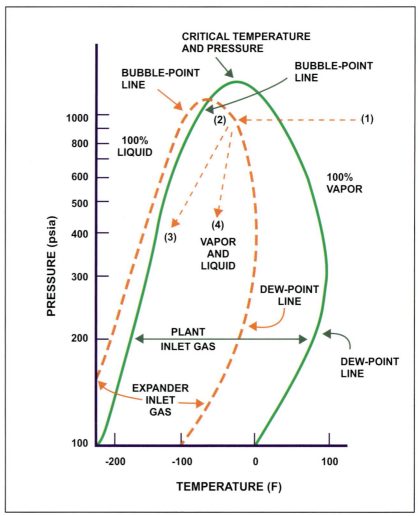

Source: *GPSA Engineering Data Book*, 12th edition

Figure 8.3   Pressure-temperature diagram for the turboexpander process

Plant Processing of Natural Gas

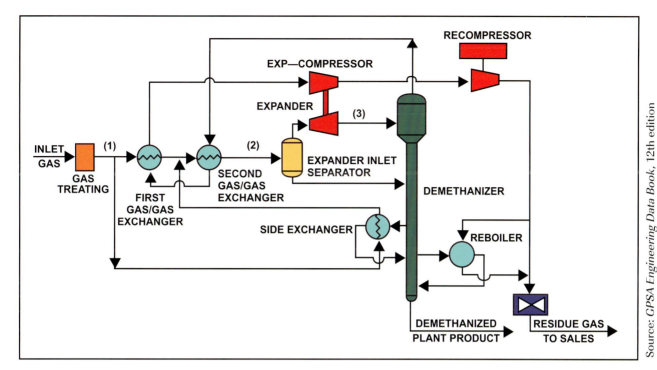

Figure 8.4   Diagram of a plant using a turboexpander process

(This is called a Joule-Thompson expansion.) Both the outlet pressure and temperature would be higher, resulting in a reduction of product recovery. The outlet pressure is higher because the expander-compressor is not available for compression. A larger compressor would allow a lower outlet pressure, but the outlet temperature would still be higher and reduce product recovery.

The gas must be properly prepared for processing. The most difficult problem is usually the removal of water. A moisture analyzer can be used to check the water content of the gas entering the heat exchanger and other plant equipment. Close monitoring of the inlet gas pressure drop as it flows through the various heat exchangers is also necessary. An increase in the inlet gas pressure drop indicates obstruction of the inlet gas flow path due to freezing (hydrate formation) or foreign matter, such as desiccant particles from the dehydration unit.

Proper product demethanization depends primarily on the demethanizer bottom temperature. The correct bottom temperature depends on the product recovery level, the inlet gas heavy hydrocarbons content, and the demethanizer pressure. Low ethane recovery requires a higher bottom temperature. If the expander-compressor or one of the recompressors is down, the demethanizer pressure must be increased. Higher bottom temperature is required at a higher operating pressure.

The most important temperature for high product recovery is the vapor temperature to the expander. A lower temperature to the expander increases product recovery. However, the expander process also condenses a considerable quantity of methane while condensing the recovery product (ethane, propane, butanes, and natural gasoline). Too much methane condensation can overload the demethanizer. A lower temperature to the expander increases methane condensation. Therefore, the vapor temperature to the expander should be as low as possible but not overload the demethanizer.

In the 1970s, the increase in turboexpander plants was largely driven by:

- a significant increase in ethane value as a liquid compared to ethane Btu value in the gas phase;
- technological improvements in the manufacture of practical mechanical turboexpander designs suitable for continuous operation in a variety of operating conditions;
- development of compact, efficient, and relatively inexpensive plate-fin or brazed aluminum heat exchangers at the same time as reliable turboexpanders.

During the 1970s, it was also recognized that recompression, generally accounting for 25%–50% of a gas plant's cost, could be reduced by increasing the operating pressure of the demethanizer. The limitation of this concept is illustrated in figure 8.5.

Figure 8.5 shows that as the pressure approaches critical pressure, the relative volatility (vertical dashed lines) decreases, making fractionation more difficult. Moreover, higher demethanizer pressure reduces the expansion ratio across the expander, and the temperature profile in the column rises, making the heat integration into the process more difficult.

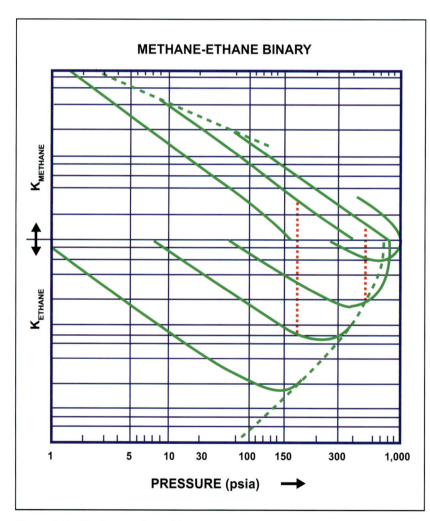

Source: *Oil & Gas Journal, "Petroleum in the 21st Century," 1999*

Figure 8.5   Methane-ethane binary

To compensate for the reduction in relative volatility, ongoing advancements have concentrated on improving the reflux to the demethanizer rectification section. For instance, liquid recovery might be improved by the addition of a reflux condenser to the demethanizer (Gulsby, 1981). Refrigeration for the reflux condenser is provided by the expander discharge vapor after being separated from condensed liquid in the demethanizer.

The most effective and widely used method is a split-vapor process developed independently for a project in Australia (White-Stevens, 1986) and also by the Ortloff Corporation in Midland, Texas (Campbell, 1979). This process, referred to as the *gas sub-cooled process (GSP)*, uses a fraction of the vapor from the cold separator as the top reflux to the demethanizer, after substantial condensation and subcooling (fig. 8.6). The main fraction, typically in a range of 60%–70%, is subjected to turboexpansion as usual.

Despite less flow being expanded by the turboexpander, this higher and colder reflux flow results in an improved ethane recovery, even at a higher column pressure. This method reduces the recompression horsepower requirements. Another advantage is the reduction of the risk of CO_2 freezing in the demethanizer.

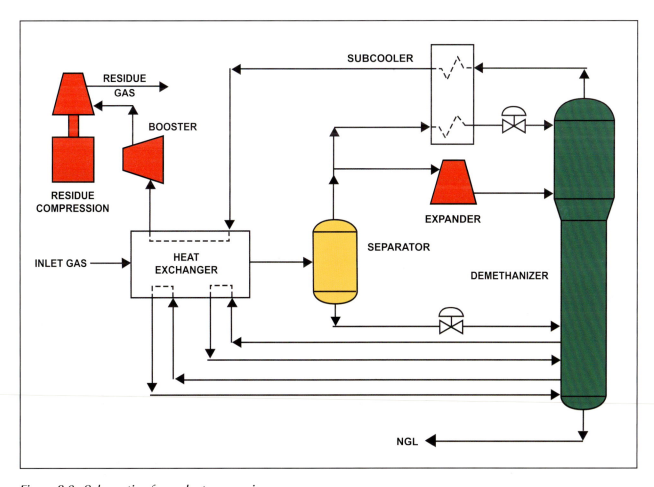

Figure 8.6   Schematic of gas plant processing

Propane-Recovery Process

Because of equilibrium constraints, achievable propane recovery during ethane rejection in the GSP process is limited by the propane content in the top reflux to the demethanizer.

For high propane recovery (98+%), a two-column scheme similar to figure 8.7 is used. In this process, the overhead vapor from the second column (a *deethanizer*, for example) is condensed and recycled to the top of the demethanizer.

Typically, this scheme is used in a two-column arrangement with the demethanizer comprising only the top rectification section similar to an absorber (Buck, 1986). The absorber bottom liquids are pumped to the top of the deethanizer. The top reflux is leaner in propane, and enhances the recovery efficiency of propane and heavier hydrocarbons.

Improvements to the two-column arrangement are made by thermally integrating the two columns to enhance the effectiveness of the reboiler heat and refrigeration requirements (Yao, 1998). Incorporating a stripper section allows undesirable light components to be stripped off the liquid feed to the deethanizer.

The result is that energy requirements for the deethanizer can be substantially reduced. In addition, the overhead deethanizer vapor is almost pure ethane and free of propane, increasing propane recovery efficiency. Propane recovery in excess of 99% can be achieved, even with a relatively lower expansion ratio.

The second tower, originally designed and operated as a deethanizer, can also produce ethane product when the plant is switched to an ethane recovery operation.

Figure 8.7   Deethanizer overhead recycle process

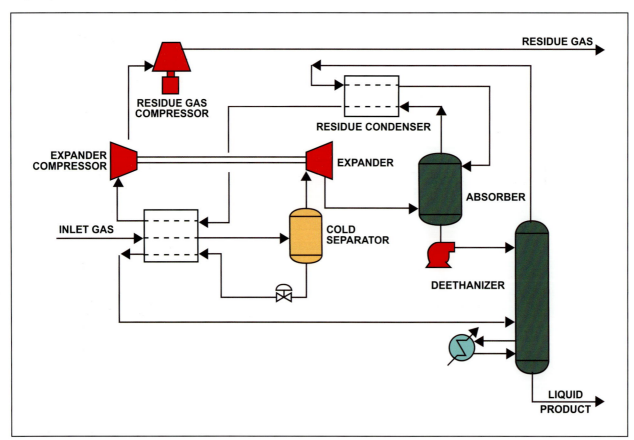

Source: *Oil & Gas Journal*, "Petroleum in the 21st Century," 1999

Ethane-Recovery Process

The recovery level possible in the split-vapor process is limited by the composition of the vapor stream used for the top reflux to the demethanizer. Using a leaner reflux attempts to overcome this deficiency. In some processes, recycling part of the residue gas as the top reflux to the demethanizer can enhance ethane recovery (fig. 8.8).

Because the refrigeration from the residue gas is not cold enough to liquefy the slip recycle stream, a compressor operating at cryogenic temperatures is used to raise the recycle stream pressure and make the main cold residue gas condense.

In this method, the sub-cooled split-vapor stream is inserted into the rectification section, providing the bulk reflux to the expander discharge vapor (Montgomery, 1989). The small amount of ethane remaining in the up-flowing vapor subsequently contacts with the essentially ethane-free top reflux and is retained in the down-flowing liquid.

The result is a smaller recycle flow and less compression required for a specific recovery level. High ethane recovery in excess of 98% is technically achievable with this method. However, the cryogenic compressor is an expensive piece of equipment.

An alternate method in which warm recycled residue gas is taken from the residue-gas compressor has also been used (Aghili, 1987; Campbell, *et al.* 1996). The compression requirements for the recycle residue gas are achieved by the main residue-gas compressor, eliminating the need for a dedicated booster compressor and reducing capital investment. This method also produces very high ethane recovery.

Figure 8.8   Residue gas recycle process

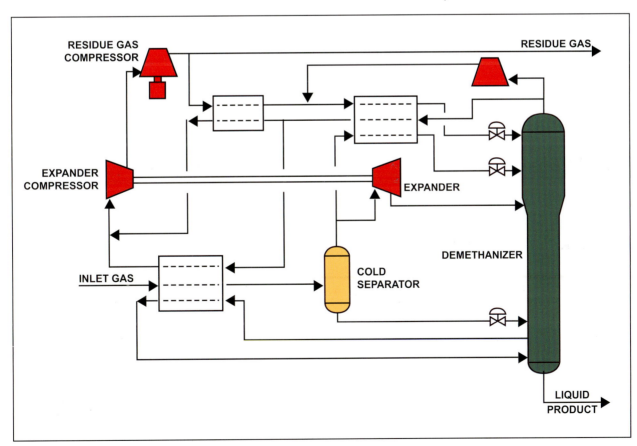

Source: *Oil & Gas Journal,* "Petroleum in the 21st Century," 1999

The recovery of hydrocarbon liquids (governed by phase equilibrium at the column top stage) can be enhanced by a higher, colder, or leaner top reflux. Many patents have been issued describing alternative schemes for improving the reflux for demethanizers and deethanizers. One unique self-refrigeration scheme has been patented that enhances the stripping (bottom) section of the towers. The reader is referred to several good references at the end of this chapter that provide additional information on these process schemes (Lee, 1999; Pitman, 1998; Elliot 1996).

TURBOEXPANDERS

Turboexpander use in the natural gas and other hydrocarbon industries began in the 1960s and has become universally accepted. The size of turboexpanders used in the hydrocarbon industry range from approximately 250 horsepower and up. A few have reached the 20,000 horsepower range. Turboexpanders are widely used in the hydrocarbon industry today for the recovery of ethane and propane from natural gas. They are also used in the recovery of power from the expanding streams during the processing of helium, carbon dioxide, and hydrogen. Turboexpanders are also used in the liquefaction of helium, hydrogen, and natural gas (Swearingen, 1971).

When used to drive a compressor, the turboexpander has as its fundamental purpose the generation of low-temperature refrigeration required by the cryogenic process. Figure 8.9 shows a packaged turboexpander-compressor.

In most petroleum processes, a Btu subtracted by refrigeration at low temperature (the opposite of heating) is seven times more valuable in terms of work expenditure than a Btu of added heat. Therefore, refrigeration processes must be completely and carefully insulated to prevent the heat gains acceptable in most petroleum processes to prevent heat losses at the higher temperatures. Because of the efficiency-temperature relationship, the turboexpander should be located in the system at the lowest temperature position possible.

The efficiency of the turboexpander is equal to, or better than, mechanical compressor drives at temperatures below –75°F (fig. 8.10). The turboexpander is intimately integrated with each component responsive to the characteristics of each of the system's units.

J/N EC3.5-578, S/N 0301, MADE IN USA

Figure 8.9   A 3.5 turboexpander-compressor used to process offshore gas from the Gulf of Mexico

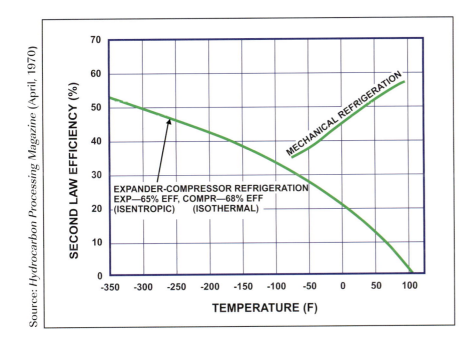

Source: *Hydrocarbon Processing Magazine* (April, 1970)

Figure 8.10   Efficiency of turboexpansion cooling

An important system design consideration is the control of the turbo-expander and the absorption of the turboexpander power. If power from the turboexpander can be absorbed by a compressor where the mass flow is proportional to the turboexpander flow, control is simplified. As illustrated previously in figure 8.3, this method results in power absorption at all flow rates at speeds that closely approach the optimum efficiency speeds. In this arrangement, speed control is not normally required because the speed will adjust within a reasonable range due to the power balance between the compressor and the expander.

A radial-reaction type of turbine achieves the greatest efficiency by using a variable nozzle assembly surrounding the rotor. A variable nozzle assembly surrounds the rotor. The nozzle blades can be tilted to vary the clearance between them so the available flow area is the nozzles that inject the gas into the rotor (fig. 8.11).

Courtesy of Cryostar SAS

Figure 8.11   A radial-reaction turbine showing nozzle blades

This is effective in controlling the flow in a turboexpander over a wide range, even above design flow. The fixed losses in the turboexpander quickly reduce overall efficiency at low flows because these losses must be prorated against a lower flow. There is a combination of rotor and nozzle that gives a high efficiency over a wide range that is commercially available.

Operating the turboexpander at the lowest possible temperature to attain process efficiency must be done on a gas stream that partially condenses as it passes through the turboexpander. This places special requirements on the turboexpander. There are turboexpanders available that claim no efficiency loss for any amount of liquid condensed.

In addition to wide flow variation capability, some applications require somewhat wide pressure variation. In these situations, a loading device consisting of a compressor or turbulence device, the speed of which responds to some degree of the applied power, can be used. The speed will closely follow the speed corresponding to optimum efficiency (fig. 8.12).

Courtesy of Cryostar SAS

Figure 8.12   Turboexpander

The turboexpander is a small part of the investment in a cryogenic plant, yet its performance in terms of efficiency and reliability are important to the success and profitability of the process. The design, startup, and operation of each installation must carefully avoid exposing the turboexpander to severe operating conditions. Otherwise, the plant might shut down or have impaired performance.

Usually, the turboexpander is started and operated at reduced speed for a period of time. This is done for a variety of reasons, such as cooling down the plant or bringing other units on stream. The turboexpander might not reach full speed for quite some time. If the turboexpander must be capable of variable-speed operation, it cannot be a flexible shaft design. Rather, it must be a stiff-shaft design to avoid shaft critical speed (fig. 8.13).

This also applies to the bearings used. They must not be of a design in which the oil film is critical within the operating speed. The bearings should be designed to carry the heavy loads created by the imbalance caused by freeze-up or erosion. Heavy-duty bearings have the added quality of avoiding oil-film *resonance* or vibration. These bearings required the shaft to be of large diameter, which, in combination with rigid bearings, maintains excellent seal alignment.

Some turboexpanders are built with tilting-shoe bearings. The purpose of tilting shoe bearings is to suppress *oil whirl,* which occurs when oil draws into a wedge rather than flows smoothly around the bearings. Oil whirl can result in *oil whip,* a dangerous condition where the rotor uses up the entire bearing clearance and is in direct metal-to-metal contact. This phenomenon occurs at roughly twice the frequency of the oil-film resonance, so if the oil-film resonance does not occur, then oil whirl is not a problem.

Courtesy of Cryostar SAS

Figure 8.13   Wheel shaft

Over three decades ago, active magnetic bearings were introduced to study the feasibility of replacing conventional journal or ball bearings (fig. 8.14). Eliminating the problems associated with lubrication made their use attractive in a variety of applications for rotating machinery.

Magnetic bearings used in the industry are a valuable machine element in which load, size, stiffness, temperature, precision, speed, losses, and dynamics are variables. Magnetic bearings have been applied to a variety of rotating equipment including compressors, steam and gas turbines, pumps, fans, centrifuges, and turboexpander compressors. While the usual limitations remain, magnetic bearings with high loads (rotors with a mass of 50 tons) are considered to be state-of-the-art. In the gas processing and hydrocarbon industry, turboexpander compressor machinery is operating with over 98% reliability.

There are advantages and disadvantages to magnetic bearings:

Advantages:

- Increased mechanical efficiency from 1% to 2%
- No oil system reduces installation, maintenance, and safety costs
- Reduced weight and size of skid
- Bearings can be cooled with process gas
- Optimize performance

Disadvantages:

- Bearings are large to provide adequate force
- Stiffness is lower than conventional oil bearing

A potentially serious problem with turboexpanders is the possibility of resonance in the rotor that is excited by a frequency equal to the running speed multiplied by the number of blades in the rotor. This is a frequency on the order of several thousand cycles per second. If such resonance is encountered during operation, it will destroy the rotor in a short time. Therefore, the rotor must be designed so the operating range is free of all such resonance frequencies.

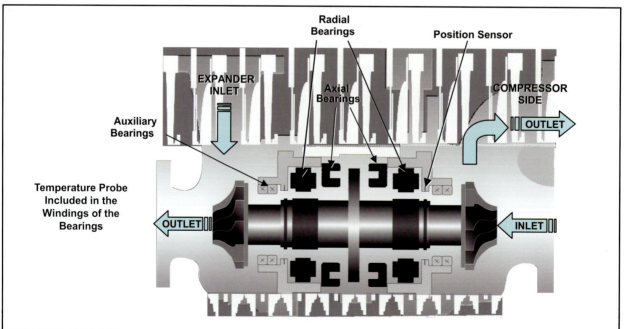

Figure 8.14 *Active magnetic bearings for typical ABM turboexpanders*

Plant Processing of Natural Gas

CRYOGENICS

Cryogenic gas processing requires specialized types of heat exchangers. Below –50°F, the common shell-and-tube heat exchanger becomes uneconomical in comparison to the specialized class of exchangers known as plate fin, or core exchangers. Most cryogenic exchangers are made of aluminum because of the increasing tensile strength as the temperature drops. From a cryogenic standpoint, aluminum has two poor qualities: large thermal expansion and fatigue cracking from repeated stresses. A core exchanger is rated by the cubic foot of core and is built as a rectangular solid with external headers.

A typical plate-fin exchanger will provide 300–400 square feet of heat transfer surface per cubic foot of exchanger volume, or 6–8 times the surface density of comparable shell-and-tube exchangers. Moreover, the plate-fin exchanger provides approximately 25 times more surface per pound of equipment than comparable shell-and-tube exchangers.

The design lends itself to a variety of streams with a large extended surface area per cubic foot of core. The core is constructed of an aluminum plate, an aluminum corrugated section, a second aluminum plate, a second corrugated section, and so on until the desired core is built up (fig. 8.15).

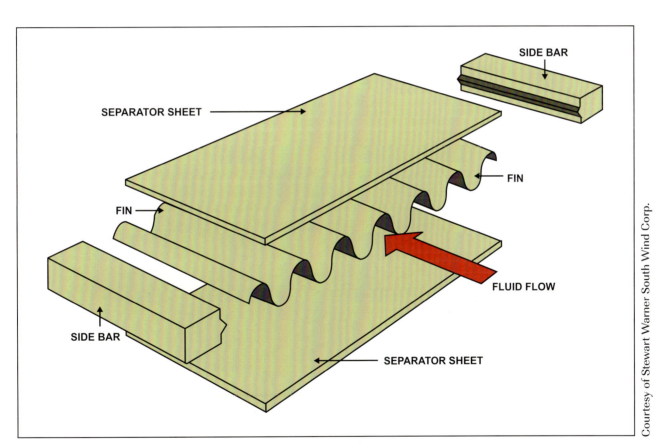

Figure 8.15   Core of an aluminum plate

Courtesy of Stewart Warner South Wind Corp.

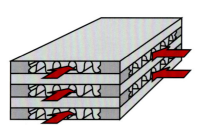

CROSS-FLOW
The corrugations (fins) in alternate layers are arranged so that the flow channels are 90° with respect to each other.

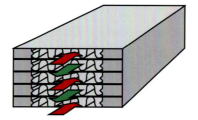

COUNTER-FLOW
The corrugations (fins) are arranged so that all the flow channels are parallel. The fluid streams in alternate layers flow in opposite directions.

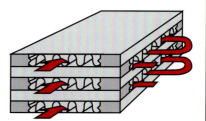

MULTI-PASS-FLOW
The flow channels are positioned as in the cross-flow pattern with selected passages separated and sealed to provide a multi-pass arrangement.

Figure 8.16   Corrugated fin flow patterns

The corrugated areas can be either parallel flow or cross flow. A great variety of exchangers can be built up by a proper selection of headers and side bars to seal the core where desired (fig. 8.16).

This type of exchanger is compact and has very high performance. The exchangers are rated with a 2°F–3°F approach temperature and little pressure drop. The heat transfer with this type of exchanger is efficient, compact, and less expensive than the alternatives: stainless steel shell-and-tube exchangers (figs. 8.17 and 8.18).

Figure 8.17   Cores of plate-fin heat exchanger (PFHE)

Plant Processing of Natural Gas

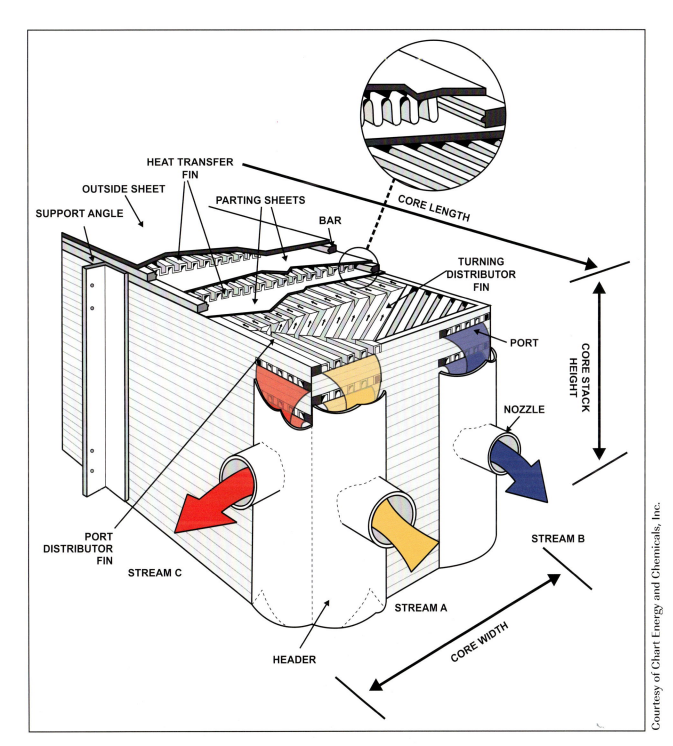

Figure 8.18   Components of a brazed aluminum heat exchanger

REFERENCES

Aghili, H.K., U.S. Patent No. 4,687,499, McDermott (1987).

Campbell, R.E., and Wilkinson, J.D., U.S. Patent No. 4,157,904, Ortloff (1979).

Campbell, R.E., Wilkinson, J.D., and Hudson, H.M., U.S. Patent No. 4,889,545, Ortloff (1989).

Campbell, R. E., Wilkinson, J. D., and Hudson, H. M., U.S. Patent No. 5,568,737, Ortloff (1996).

Elliot, D.G., Chen, J.J., Brown, T.S., Sloan, E.D., Kidnay, A.J., "The Economic Impact of Fluid   Properties Research on Expander Plants," Fluid Phase Equilibria, Vol. 116, Elsevier Inc., Burlington, Massachusetts (1996).

GPSA Engineering Data Book, 12th edition, Gas Processors Suppliers Association, Tulsa, Oklahoma (2004).

Gulsby, J.G., U.S. Patent No. 4,251,249, Randall (1981).

Huebel, R.R., U.S. Patent No. 4,519,824, Randall (1985).

Lee, R.J., Yao, J., Elliot, D.G., "Flexibility, Efficiency to Characterize Gas-Processing.

"Technologies in the Next Century," *Oil & Gas Journal*, Petroleum in the 21st Century, December 12, 1999.

Lee, R.J., Yao, J., Chen, J.J., and Elliot, D.G., "Lean Reflux Process for High Recovery of Ethane and Heavier Components," U.S. Patent No. 6,244,070 B1 (2002).

Montgomery, G.J., U.S. Patent No. 4,851,020, McDermott (1989).

Pitman, R.N., Hudson, H.M., Wilkinson, J.D., Cuellar, K.T., "Next Generation Processes for NGL/LPG Recovery," 77th Annual GPA Convention, Dallas, Texas (1998).

Swearingen, J., "Engineer's Guide to Turboexpanders," *Hydrocarbon Processers Magazine* issue of April 1970.

Swearingen. J., "Tuboexpanders and Processes that Use Them," proceedings of the 64th annual AIChE meeting, Nov. 29, 1971, San Francisco, California.

White-Stevens, D.T., Elliot, D.G., "Massive Cooper Basin Liquids Project in Australia Meets   Design and Startup Targets," *Oil & Gas Journal* (January 20, 1986).

Yao, J., Chen, J.J., Lee, R.J., Elliot, D.G., "Improved Propane Recovery Methods," U.S. Patent Application No. 09/209,931 (1998).

Yao, J., Chen, J.J., and Elliot, D.G., "Enhanced NGL Recovery Processes," U.S. Patent No. 5,992,175 (1997).

* Note: The authors are grateful to the *Oil & Gas Journal* for permission to use material previously published in their magazine.

FRACTIONATION

The basis for most hydrocarbon phase separations is the equilibrium flash process. A flash process gives a "sloppy" or imprecise separation of the components in a mixture. The liquid and vapor phases of the flash process still contain all the components that were present in the feed gas. To better separate the feed components, a process called fractionation, or distillation, was developed. Fractionation is possible when the components to be separated have different boiling points. The higher the difference in the boiling points, the easier it is to separate the components.

During fractionation, a mixture is separated into individual components or groups of components. Fractionation is a *countercurrent operation* in which vapor mixtures are repeatedly brought in contact with liquid mixtures having similar composition as the vapors. The liquids are at their *bubble points* and the vapors are at their dew points. Bubble point is the temperature at which the first bubble of gas forms in liquid. Part of the vapor condenses, and part of the liquid vaporizes during each contact. The vapor becomes richer in the lighter or lower boiling components, and the liquid becomes richer in the heavier or higher boiling components.

A fractionating column can be viewed as a combination of absorption and stripping columns. Figure 9.1 is a schematic diagram of a fractionation column with the associated and peripheral equipment.

The cooling in a condenser is done either by air, cooling water, or a refrigerant such as propane. The column pressure normally determines the cooling medium that is used. Reboiler heat is provided by steam, hot oil, hot medium fluids, hot compressor discharge gas, or a hot process stream.

The number of trays or the height of the packed section in a fractionation column depends upon the number of vapor/liquid contacts required to make the desired separation. Fractionation columns use *valve trays* with *downcomers* or pipes that move the liquid from one tray to the one below. The valves open either partially or fully by vapor flowing through the tray. A *weir* maintains liquid level on the tray. Liquid flows across the tray, over the weir, and through a downcomer or down pipe to the tray below. Large-diameter columns might have two or four liquid flow-passes on each tray. Figure 9.2 shows flow through vapor passages on a tray in a fractionation column.

9
Fractionation and Liquid Treating

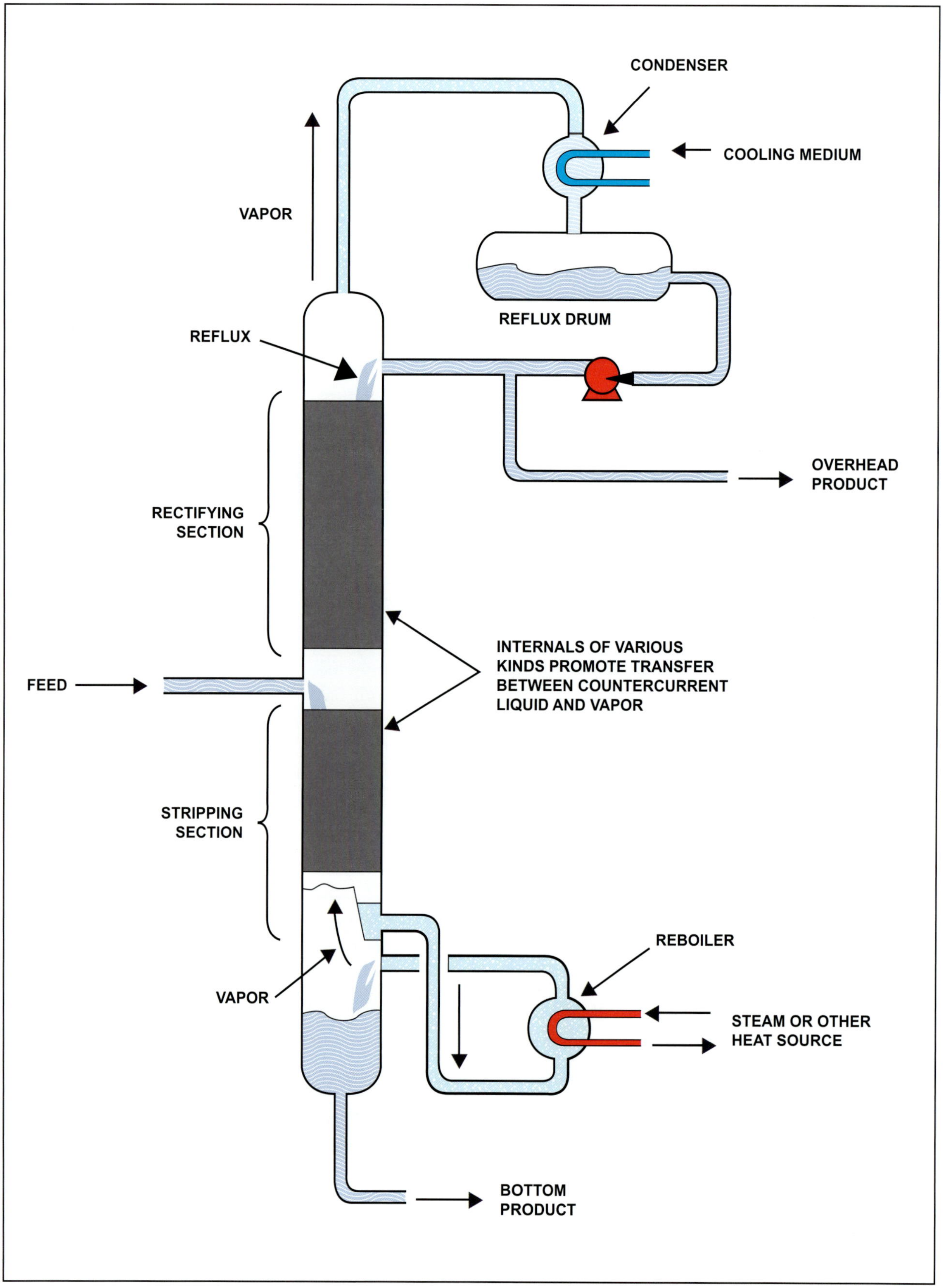

Figure 9.1   Diagram of a fractionation column

Source: *GPSA Engineering Data Book*, 12th edition

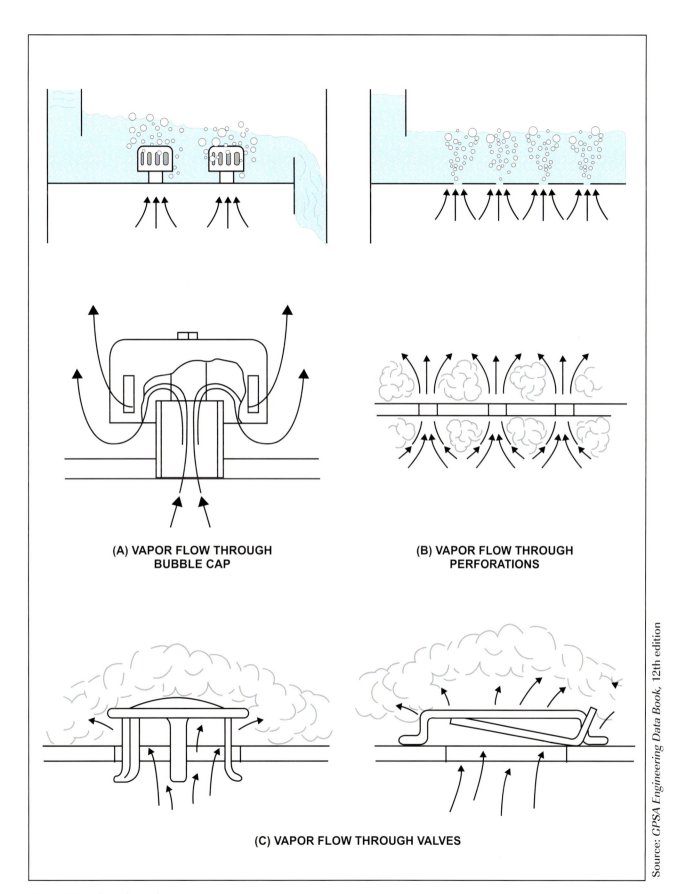

(A) VAPOR FLOW THROUGH BUBBLE CAP

(B) VAPOR FLOW THROUGH PERFORATIONS

(C) VAPOR FLOW THROUGH VALVES

Source: *GPSA Engineering Data Book*, 12th edition

Figure 9.2   Flow through vapor passages

PACKED COLUMNS

The majority of distillation columns in gas processing plants are equipped with trays. Packing is an alternative option to using trays. Compared to trays, the advantages of packed columns are:

- lower pressure drop through the column;
- smaller column diameter resulting in capital cost savings.

The disadvantages of packing versus trays are:

- limited turndown capability;
- possible problems with unequal liquid distribution;
- more susceptible to fouling.

Packing material might be metal, plastic, or ceramic. There are two main types of packing used in distillation columns:

- Random packing in which discrete pieces of packing are randomly dumped into a column shell (fig. 9.3).
- Structured packing in which a specific geometric configuration is achieved. This type of packing is particularly attractive for high liquid loading applications (e.g., demethanizer columns in a cryogenic NGL recovery plant) (fig. 9.4).

Figure 9.3 Various types of random packing

Courtesy of Beihai Kaite Chemical Packing Company Ltd.

NGL FRACTIONATION PLANTS

In an NGL fractionation plant, the number of fractionation columns required depends on the number of products to be made and the NGL composition. The plant is designed to make products that meet the specifications stated in the sales contracts. Fractionation columns are usually named for the top or *overhead product* that they make. For example, a deethanizer column means that the top product is predominantly ethane; a depropanizer column means that the top product is predominantly propane; and so on. Normally, a plant has a deethanizer (DeC_2), depropanizer (DeC_3), debutanizer (DeC_4), and a deisobutanizer (DIB) that is also referred to as a butane splitter (fig. 9.5).

STRUCTURED PACKING

PACKED TOWER
INTERNALS

RANDOM PACKING

HIGH-PERFORMANCE
TRAYS

CONVENTIONAL
TRAYS

Figure 9.4   Tower with various packing materials including structured packing

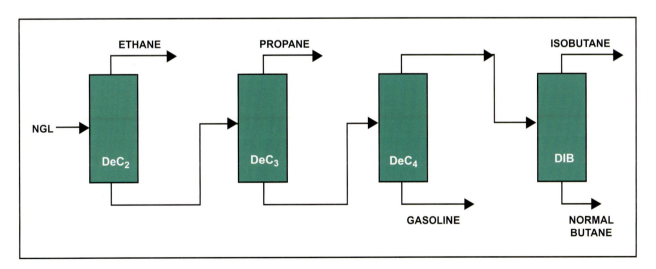

Figure 9.5   Example of a four-column fractionation plant

Fractionation and Liquid Treating

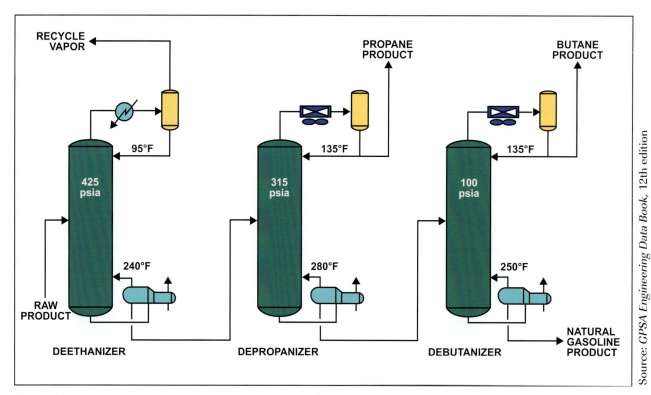

Source: GPSA Engineering Data Book, 12th edition

Figure 9.6   Example of a fractionation plant used to produce three products

An example of a fractionation plant used to produce three products is shown in figure 9.6.

Deethanizer (DeC$_2$) Column

The first step in fractionating NGL is to separate ethane from propane. In DeC$_2$, the overhead is the ethane product and the bottom products contain propane and heavier components. In operating this column, the key variables are:

- *Top*: the amount of propane and heavier components (C$_3$+) left in the ethane product. This is done by controlling the reflux rate. The maximum C$_3$+ components left in the ethane product depend on product specifications and on the economic value difference between ethane and propane products and energy prices.

- *Bottom*: the amount of ethane left in the bottom product. This is done by controlling the reboiler temperature. Because ethane is worth less than propane (C$_3$), the amount of ethane in the bottoms is maximized. The maximum amount of ethane in propane product is dictated by the vapor pressure specification of the C$_3$ product, because all of the ethane will go out with the propane product in the depropanizer.

Plant Processing of Natural Gas

Depropanizer (DeC$_3$) Column

In this column, a C$_3$ product goes overhead while isobutane (i-C$_4$) and heavier components are produced as bottom product. In operating this column, the key variables are:

- *Top*: the amount of i-C$_4$+ components left in the propane product. This is done by controlling the reflux flow rate. All the C$_2$ left in the DeC$_3$ feed goes out with the propane product. The maximum amount of i-C$_4$+ components left in the C$_3$ product depends on its specifications and on the economic value difference between C$_3$/i-C$_4$ products and energy prices.

- *Bottom*: residual amount of propane left in the bottom product. This is done by controlling the reboiler temperature. Any C$_3$ left in the bottoms ends up in the i-C$_4$ product in the butane splitter. Normally, C$_3$ is worth less than i-C$_4$. Therefore, the goal is to maximize the amount of C$_3$ left in the bottom product consistent with meeting i-C$_4$ product specifications.

Debutanizer (DeC$_4$) Column

The bottoms from the DeC$_3$ flow to the DeC$_4$ column as feed. In this column, normal butane (n-C$_4$) and lighter components (C$_3$ and i-C$_4$) go overhead while C$_5$+ components (gasoline) go out as bottom products. In operating this column, the key variables are:

- *Top*: the amount of i-C$_5$+ components left in the overhead mixed butanes stream. This is done by controlling the reflux flow rate. The C$_5$+ components, left in the mixed butanes stream, end up in normal butane product in the butane splitter. Because C$_5$+ components are usually more valuable than a normal butane product, the goal is to control their amount based on normal butane product specifications and the economic value difference between gasoline/n-C$_4$ product and energy prices.

- *Bottom*: the residual butanes in the bottom product. This is done by controlling the reboiler temperature. The n-C$_4$ is usually worth less than a gasoline product. Therefore, the amount of n-C$_4$ left in gasoline is maximized to meet the gasoline product specifications, which include vapor pressure and/or maximum n-C$_4$ fraction.

Deisobutanizer (DIB) or Butane Splitter Column

The mixed butanes from the DeC$_4$ column flow to the DIB as feed. In the DIB, i-C$_4$ product goes overhead whereas normal butane product is the bottoms product. In operating this column, the key variables are:

- *Top*: amount of the n-C$_4$ left in the isobutane product. This is done by controlling the column reflux flow rate. Any C$_3$ left in the DeC$_3$ bottoms goes out in this stream. Usually, i-C$_4$ is more valuable than n-C$_4$ product. Therefore, the amount of n-C$_4$ in the overhead is maximized to meet product specifications.

- *Bottom*: the residual i-C$_4$ in the bottom n-C$_4$ product. This is done by regulating the reboiler heat input. The i-C$_4$ is usually more valuable than the n-C$_4$ product. The residual i-C$_4$ amount in the n-C$_4$ product is minimized, consistent with product specifications and the value difference between i-C$_4$/n-C$_4$ products and energy prices.

PRODUCT SPECIFICATIONS

The specifications for propane, butanes, and butane-propane mix (LPG) are detailed in GPA Publication 2140, "Specification and Test Methods for LP-Gas." The specifications for natural gasoline are given in GPA Publication 3132, "Natural Gasoline Specifications and Test Methods." Various grades of gasoline are established based upon Reid Vapor Pressures (10–34 psia) and the percentage of product evaporated at 140°F is 25%–85%.

Industry standards do not exist for ethane, ethane-propane mixes (EP mix), isobutane, and normal butane products. These specifications are developed on a plant-by-plant and sale-by-sale basis. Table 9.1 shows representative product specifications for an NGL fractionation plant located in Southern Louisiana.

Table 9.1
Product Specifications for a Southern Louisiana Fractionation Plant

Purity Ethane		
CO_2	500 ppm by wt	(max)
Methane	5% mole	(max)
Ethane	90% mole	(min)
Propane	5% mole	(max)
Butanes+	0.25% mole	(max)
Moisture	76 ppm by wt	(max)
Free O_2	5 ppm by wt	(max)
Total Olefins	3% mole	(max)
Total Sulfur	5 ppm by wt	(max)
Contaminants		
• *Ammonia*	1 ppm by wt	(max)
• *Arsine*	10 ppm by wt	(max)
• *Phosphene*	10 ppm by wt	(max)
Propane, HD-5		
Reid Vapor Pressure	208 psig	(max)
Propane	90.0% vol	(min)
Propylene	5.0% vol	(max)
Butanes+	2.5% vol	(max)
Total Sulfur	23 ppm by wt	(max)
COS	15 ppm by wt	(max)
Isobutane		
Propane	3.0% vol	(max)
I-Butane	95% vol	(min)
N-Butanes+	5.0% vol	(max)
Volatile Sulfur	140 ppm wt	(max)
Normal Butane		
i-Butane	5.0% vol	(max)
N-Butane	95% vol	(min)
Pentanes+	2.0% vol	(max)
Propane	3.0% vol	(max)
Volatile Sulfur	140 ppm by wt	(max)
Gasoline		
Vapor Pressure (at 100°F)	13.0 psia	(max)
N-Butane	2.0% vol	(max)

MONITORING FRACTIONATION PLANT OPERATION

To maximize net income from a fractionation plant, the key variables are monitored to ensure that they are close to the plant's economic optimum (fig. 9.7). The key variables are:

- Product purity
- Column pressure
- Reflux rate
- Reboiler duty
- Key temperatures
- Reboiler/condenser performance (the heat transfer coefficient)

Courtesy of Dominion

Figure 9.7   Fractionation plant

The columns in a fractionating plant are designed for the worst summer cooling water or air temperatures. However, for most of the year, the actual temperatures are significantly lower than design specifications. This allows leeway for the lowering of the column operating pressures until a mechanical constraint, such as column flooding, is encountered. Because at lower pressures they boil at lower temperatures, hydrocarbons have greater volatility differences at low pressures and are, therefore, easier to separate. The column bottom temperature can be reduced because hydrocarbons at lower pressures will boil at a lower temperature, saving process heat. Two examples of heat savings are given next.

Depropanizer example:

Design pressure 240 psig and the reflux is 114°F. If the pressure is reduced to 180 psig when reflux T is 91°F, the reboiler duty will decrease by 16%.

Debutanizer example:

Design pressure is 87 psig and reflux is 120°F. If the pressure is reduced to 54 psig with reflux T is 88°F, the reboiler duty will decrease by 18%.

POSSIBLE OPERATING PROBLEMS

The most common and frequent problems with fractionators are the variations in feed flow rates and the feed compositions. Most fractionation plants have a feed surge tank located upstream of the fractionating columns. In most modern fractionation plants, computer management controls are used to keep constant feed rates to the columns while the liquid level in the surge tank is allowed to fluctuate within a wide band. Only when the level gets too close to the limits are the feed rates to the columns changed. When a plant is running close to its capacity, even small changes in flow rates and/or compositions can result in off-specification products and/or column flooding unless operating changes are quickly made. Computer management controls are designed to make these quick and appropriate process changes.

A fractionation column will flood when its maximum capacity is exceeded. A column floods when it has too much:

- feed;

- reflux or too low a reflux temperature, or a combination of the two; and/or

- bottom heat input.

Flooding destroys separation efficiency. When flooding occurs, reflux liquid is entrained (trapped) in the rising vapor and leaves the column overhead. If pure quality products were being produced prior to flooding, the column load should be reduced by reducing the reflux rate. If prior to flooding, product purity was just sufficient to meet sales specs, the tower load should be reduced by decreasing the feed flow rate to the column.

NGL PRODUCT TREATING

Even if the inlet gas is treated, some CO_2, H_2S, and other sulfur compounds (mercaptans, COS, and CS_2) might still remain in the NGLs recovered in a gas plant. When the recovered NGLs are transported via pipelines, they must meet CO_2, H_2S, elemental sulfur, and COS specifications. During NGL fractionation, all CO_2 and most of the H_2S end up in the ethane product; COS ends up in the propane product while the other sulfur species distribute in other fractionated products (Harryman and Smith, 1994). Ethane product is sometimes treated in the gas phase to remove CO_2 and H_2S. Gas-treating processes are covered in Chapter 4, Hydrocarbon Treating.

The process selected to treat NGLs or any of the fractionated products or their mixtures, depends on whether only trace amounts of H_2S and CO_2 are to be removed, or if the removal of other sulfur compounds, like mercaptans and COS, is also required. In general, NGL or product-treating processes can be classified as either liquid-liquid treating or liquid-solid treating.

Liquid-Liquid Treating

Alkanolamine, an aqueous solution of an alkanolamine (e.g., MEA, DEA, DIPA, or MDEA), is in contact with the liquid hydrocarbon in a packed bed or one or more mixer-settler stages. A packed tower usually has a minimum of 20 feet of packing. The design flow rates for the packed towers usually should not exceed approximately 20-gpm amine plus hydrocarbon liquid per square foot of cross sectional area. The number of mixer-settler stages depends on the residual acid gas components desired in the treated liquid.

For liquid-liquid treating, both a *coalescer* and a full-flow carbon filter should be used to minimize introduction of hydrocarbons into the amine regeneration system. Aqueous amines are used to remove CO_2 and/or H_2S from NGL (or fractionated streams) in a gas plant. Various technologies are available to remove COS from liquid propane streams, including amine treating (Rhinesmith, Archer, and Watson, 2001). An aqueous DEA or DIPA amine process is effective in removing COS from hydrocarbon liquids if adequate mixing and contact time are provided in a stirred vessel. There are many operating aqueous DIPA, including Shell's ADIP® process, used worldwide.

Caustic treating can be used to remove H_2S, CO_2, and mercaptans using sodium hydroxide.

Batch caustic wash is a one-stage process best suited for removing trace amounts of H_2S, methyl, and/or ethyl mercaptans from a liquid stream. The cost of caustic treating is high per gallon of product because the spent caustic has to be disposed of in an environmentally approved manner.

A regenerative caustic process normally uses a 10% aqueous caustic to remove H_2S, methyl, and ethyl mercaptans from hydrocarbons. If large concentrations of H_2S and CO_2 are present in the liquid stream, these are first removed with an amine unit to reduce caustic consumption. The sour caustic solvent is then regenerated by steam stripping in a column. The process can handle large volumes of hydrocarbons and can produce a doctor sweet product. Figure 9.8 shows a schematic of a regenerable caustic process.

The Merox™ process from UOP LLC uses a caustic solution containing a proprietary catalyst called Merox to extract mercaptans from a light hydrocarbon liquid. The sour Merox solvent is regenerated by mixing with air to oxidize mercaptans to disulfides (no odor). These disulfides are insoluble in the solvent and can easily be removed from the solution. The Merox process gives a very high degree of removal of mercaptans from a light liquid hydrocarbon stream.

The Merichem Company offers several processes for removal of Na_2S, H_2S, and/or mercaptans from light hydrocarbon liquids. In all the variations of the Merichem processes, fiber film bundles are used to achieve intimate contact between hydrocarbons and the caustic solution. A bundle is comprised of long, small-diameter metallic fibers that are placed in a pipe. The caustic solution preferentially wets the fibers, creating a large interfacial surface for contacting the caustic solution with the liquid hydrocarbon stream.

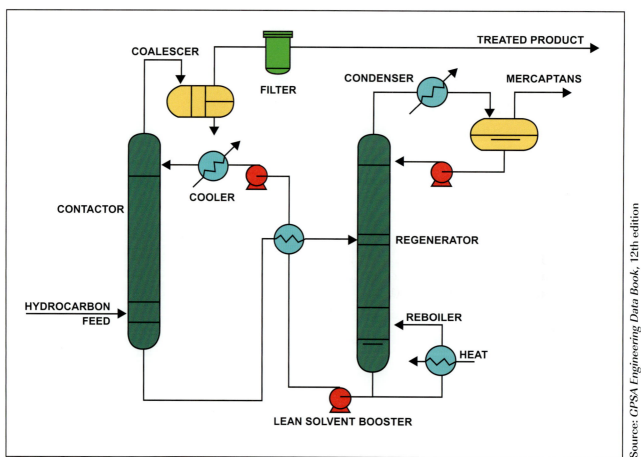

Source: GPSA Engineering Data Book, 12th edition

Figure 9.8 Schematic of a regenerable caustic process

Plant Processing of Natural Gas

Liquid-Solid Treating

In liquid-solid treating processes, a solid bed is used to remove H_2S, CO_2, mercaptans, and COS, among others, from a liquid hydrocarbon stream. The most common examples of commercial solid-liquid treating processes are the following:

- In the potassium hydroxide (KOH) process, light liquid hydrocarbons along with a small amount of water are passed over the bed of a walnut-size KOH for removal of small amounts of H_2S and mercaptans. The KOH acts as a desiccant in addition to removing the sulfur compounds. This process is suitable for removing trace amounts of H_2S. The key advantage of this process over the caustic solution is that the liquid product still meets dryness standards.

- Molecular sieves can be used to remove H_2S, COS, and mercaptans from small or large hydrocarbon streams to produce product that meets minimum specifications for sulfur compounds when subjected to 1A copper-strip or doctor sweet tests. Molecular sieves also dehydrate the product. UOP's 13X molecular sieves are effective in removing small quantities of H_2S. As with other molecular sieve applications, the sour regeneration gas must be dealt with when removing H_2S or mercaptans by this process.

- The Perco™ solid copper chloride solution utilizes copper chloride, impregnated on a stationary bed of *Fuller's earth*, to remove relatively high mercaptan levels from small gasoline streams. H_2S, sulfur, and caustic must be removed upstream of the Perco unit because they will deactivate the copper chloride. As manufactured, this material contains about 16% water and loses its reactivity below 7% wt. To maintain its activity, either a small quantity of steam is injected into the hydrocarbon feed, or a side stream is bubbled through water and returned to the main stream. Free water must also be removed to prevent extraction of copper chloride from Fuller's earth support.

REFERENCES

GPSA Engineering Data Book, 12th edition, Gas Processors Suppliers Association, Tulsa, Oklahoma (2004).

Harryman, J.M., and Smith, B., "Sulfur Compound Distribution in NGLs: Plant Test Data," GPA Section A Committee on Plant Design, P1994.15, Gas Processors Association, Tulsa, Oklahoma (1994).

Rhinesmith, R.B., Archer, P.J., and Watson, S.J., "Carbonyl Sulfide (COS) Removal from Propane," Pearl Development Company, GPA Research Report RR-175, (2001).

NITROGEN REJECTION

Nitrogen (N_2) is an inert gas found in varying amounts in natural gas reservoirs. *Nitrogen rejection* is a necessary process in maintaining a desired Btu value for sales gas and pipeline specifications. Nitrogen is added to or removed from the sales gas to adjust its heating or Btu value. Adding nitrogen lowers the Btu value, while removing nitrogen raises the Btu value. However, there is a limit to the maximum amount of N_2 or inert gases allowed. Nitrogen has an additional use in the *enhanced oil recovery* (EOR) processes and for increasing oil production through reservoir injection.

Nitrogen is removed from the feed gas at low temperatures in a *nitrogen rejection unit (NRU)* designed according to:

- Inlet gas composition
- Inlet gas pressure
- Product specifications
- Vent nitrogen
- Residue gas heating value
- Hydrocarbon recovery required

NRUs operate best under stable compositions, inlet rates, temperatures, and pressures, and must be designed to efficiently operate over a broad range of nitrogen gas compositions (5%–80%). The quantity of nitrogen in the feed gas is generally the main factor in selecting a nitrogen removal process.

NRU PROCESS SELECTION

There are four categories of processes currently available for removal of nitrogen from natural gas.

Pressure Swing Adsorption (PSA)

Pressure swing adsorption is a technology used to separate nitrogen under pressure according to its molecular characteristics and attraction to an adsorbent material at near-ambient temperatures. Special adsorptive materials are used as a molecular sieve, adsorbing the hydrocarbon components at high pressure. The process then swings to low pressure to desorb the adsorbent material. Methane is produced during the desorption step at relatively low pressure near ambient or under vacuum in some cases. It often requires pretreatment and has high capital and compression costs. The recovery of methane is generally moderate.

Cryogenic Absorption

The *cryogenic absorption process* uses chilled hydrocarbon oil to absorb the bulk of the methane and achieve a separation of nitrogen from natural gas. The absorbed methane is stripped off the oil in a regenerator and subsequently compressed back to the pipeline pressure. The need to absorb the bulk of methane requires a large oil circulation flow and equipment size. Therefore, it is most suitable for high nitrogen content streams. It has not been widely used commercially.

Membranes

In this process, membranes are used to selectively permeate methane and reject nitrogen in the gas stream. The methane recovery is generally low to moderate and is particularly suitable for a small gas stream with bulk removal of nitrogen.

Cryogenic Distillation

The *cryogenic distillation process* is done through condensation and distillation at cryogenic temperatures. It has long been used to separate nitrogen from natural gas. Because of low operating temperature, substantial pretreatment is required. It achieves high-methane recovery with a wide range of nitrogen content and is typically used for large-scale applications.

Among these four technologies, the cryogenic distillation approach has been commonly used for providing an efficient and reliable technique to upgrade natural gas by the elimination of nitrogen.

CRYOGENIC NRU PROCESSES

A nitrogen rejection facility process typically consists of four major process steps: pre-treatment, chilling, cryogenic fractionation, and recompression (fig. 10.1).

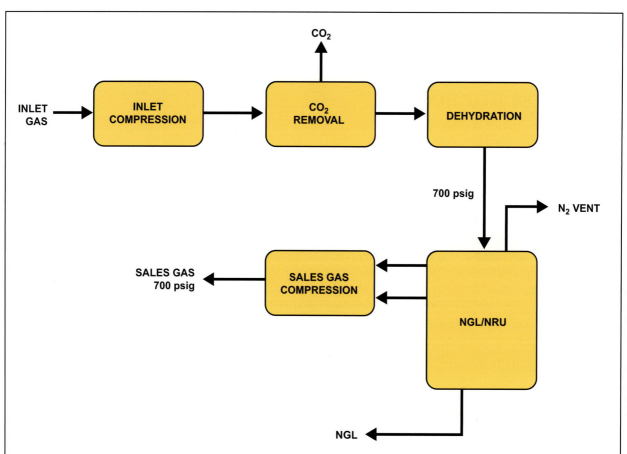

Figure 10.1   Simplified block flow schematic for a nitrogen rejection facility

Source: *GPSA Engineering Data Book*, 12th edition

Plant Processing of Natural Gas

Pretreatment

A cryogenic N_2 rejection process reaches low temperatures, typically below
$-250°F$ to $-300°F$ in the coldest portion of the plant. At those temperatures,
some components with high solidification temperatures must be reduced to
a very low concentration at the front end of the process to avoid freezing. A
list of typical components to be removed, along with their corresponding allowable concentrations, is shown in Table 10.1.

Table 10.1
Components with Allowable Concentrations in a Cryogenic N_2 Rejection Process

Property	N_2	C_1	H_2O	CO_2	H_2S	Benzene	Methanol
Molecular Weight	28	16	18	44	34	78	32
Boiling Point, °F	−320	−259	212	−109	−76.5	176	148
Freezing Point, °F	−346	−296	32.0	−69.8	−122	42.0	−144
Typical Tolerance, ppm			0.1	50-200	50-500	0.1	1.0

The allowable concentration varies with each impurity and often depends on inlet conditions, product specifications, and the process scheme
selected. For water and aromatics, such as benzene, a tolerance of less than
1 ppmv is usually recommended. The allowable concentration for carbon
dioxide can be much wider depending on the process cycle used.

Removal of other components, such as H_2S and Hg, to a low level can
also be found in the pretreatment step. H_2S is typically reduced to meet the
sales gas specification of less than 4 ppmv. Hg concentration below 1 part per
trillion (ppt) or 0.001 ppb is often required to avoid formation of free mercury,
which will attack downstream brazed aluminum exchangers and possibly
lead to catastrophic damage to the facility.

Acid gas components, such as CO_2 and H_2S, are generally removed using amine solutions. Depending upon the H_2S feed content, off-gas from the
amine unit might contain a high level of H_2S. Off-gas with a high H_2S content
must be further destroyed thermally or catalytically converted to a sulfur
element in the sulfur-recovery unit before it can be safely discharged to the
atmosphere. Alternately, acid gas can be reinjected to a dry well for economic
or environmental protection reasons. Process facilities with acid gas removal
processes add significantly to the overall cost of the plant.

After the acid gas removal, the sweet gas is most often dehydrated with a
solid *desiccant*. Molecular sieves are generally specified because of their ability
to dry the gas stream to a water dew point well below its allowable tolerance.
Any trace of mercury (Hg), if contained in the feed gas, is removed using
sulfur-impregnated activated charcoal. After drying, any fine particulates
entrained in the gas stream are captured with dust filters.

Depending on their concentrations, heavy hydrocarbons can be absorbed on charcoal or extracted at an intermediate temperature level during the chilling step. TEG commonly used for dehydration has also been commercially demonstrated for bulk removal of aromatic hydrocarbons prior to the chilling condensation step. The use of TEG upstream of the molecular sieve unit not only reduces the size of the mol sieve unit but also extends the life expectancy of the sieve.

Chilling

Following the gas pretreating step, the dry, clean gas is successively chilled to condensation temperature by heat exchange with the product streams in brazed aluminum exchangers, supplemented with external refrigeration as needed and/or by *adiabatic expansion* (JT) or isentropic expansion (turboexpander) of the feed gas. The chilled stream is now ready for fractionation where nitrogen is separated from natural gas.

All NGL components have higher condensing temperatures than methane, and all components will be liquefied in the course of the nitrogen separation process. If the NGL recovery is desired and can be efficiently integrated within the NRU facility, there can be a substantial cost saving. In a preexisting NGL recovery facility, strategically integrating the NRU and NGL facilities can also lead to substantial cost reduction by maximizing the use of the existing facilities and infrastructure.

Cryogenic Distillation

The cryogenic distillation section normally located inside the cold box is the heart of the NRU because it controls the recovery of nitrogen stream purity, hydrocarbon losses in the rejected nitrogen, and the overall thermal efficiency of the process. This operation is normally accomplished in a single-column scheme, double-column scheme, or even a triple-column scheme. Numerous variations to each scheme with differences mainly in the overall thermal integration have been made, and most of them are proprietary. Selection of an NRU cycle depends mainly on the nitrogen range of interest and product specifications, among others. Recovery of helium often contributes an important element in the overall project economics or, in some cases, becomes the main driving force for the project, particularly when the helium content exceeds 0.5 mol%. Crude helium recovery from natural gas is accompanied by nitrogen rejection in most cases and carried out in this section of the plant.

Recompression

The main force for NRU separation is provided by the pressure differential between the feed gas and the product streams. The product streams, such as sales gas and N_2, return at a pressure lower than the feed gas pressure, requiring *recompression* of sales gas to pipeline pressure for delivery. If nitrogen is needed at higher pressure, N_2 compression will also be included.

NRUs can be modified to accommodate helium recovery, but the designs are case specific.

NRU PROCESSES

Many process cycles have been developed to efficiently reject nitrogen. They can be grouped into three generic schemes:

- Single-column cycle
- Double- or dual-column cycle
- Use of a prefractionator, often referred as triple-column cycle

Although these nitrogen rejection processes perform the same task, their operating conditions might vary widely depending on many criteria, including:

- Feed gas composition variations over the life of the project
- Product values, specifications, and the use of rejected nitrogen stream (e.g., reinjection into a reservoir or vented)
- Recovery or nonrecovery of hydrocarbon liquids, and at what levels.

Generally, a single-column distillation unit with either a recycle N_2 as reflux or a heat pump cycle for providing reflux condensation allows separation at an elevated pressure (fig. 10.2). It is suitable typically when a nitrogen stream is used for reinjection. This scheme often requires an additional compressor in a heat pump cycle or for recycle N_2 as reflux.

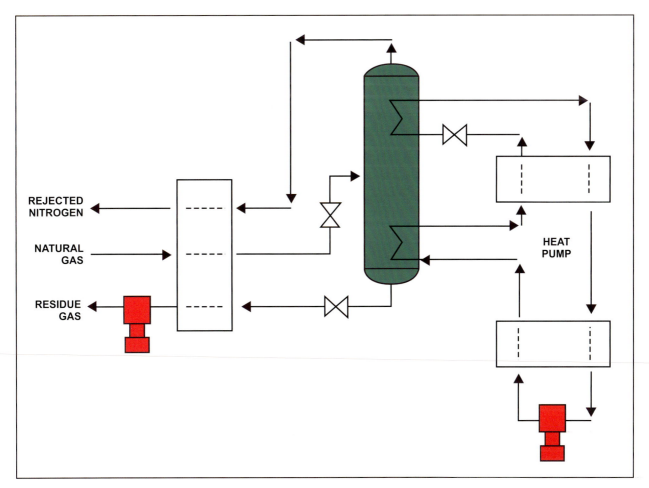

Figure 10.2   A single-column distillation unit

The double-column process relies on the pressure differential between the feed gas and the product streams to provide refrigeration for refluxing the distillation columns (figure 10.3). It produces nitrogen at basically atmospheric pressure and is best-suited for applications where nitrogen is vented. This process scheme is typically efficient for N_2 content in excess of 30%.

For a low nitrogen content case, a *prefractionator* is often employed for a bulk N_2 separation while generating an N_2-enriched feed stream prior to final separation inside a cold box. A simplified flow schematic for the prefractionator scheme is illustrated in figure 10.4.

The prefractionator is equipped with a reboiler where nitrogen is stripped off the bottom hydrocarbon liquid. This allows most of the feed natural gas to be separated and recovered from the bottom of the prefractionator at an elevated pressure of around 400 psia. This liquid hydrocarbon stream is then vaporized at the maximum pressure possible to reduce the recompression horsepower requirements after recovering its refrigeration for feed gas cooling.

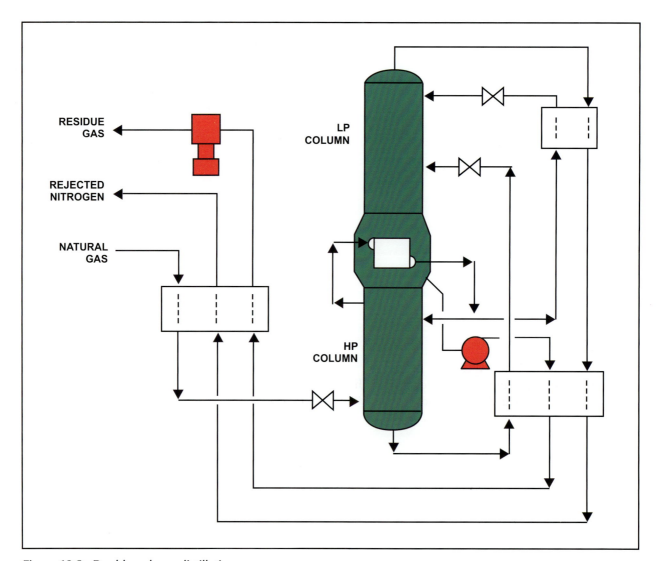

Figure 10.3   Double-column distillation process

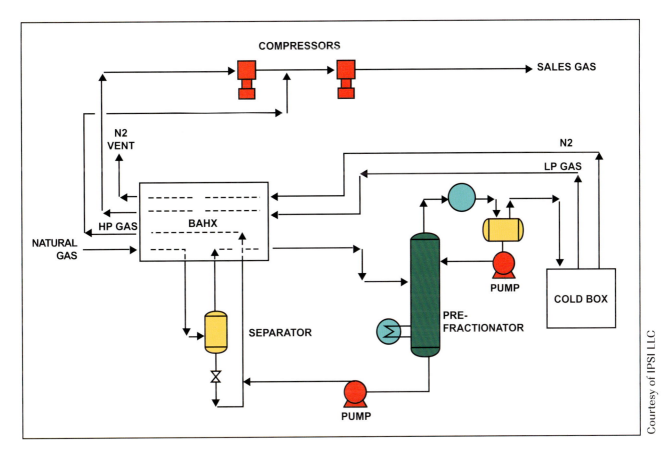

Figure 10.4 *Prefractionation scheme*

Most prefractionators are equipped with a reflux condenser to efficiently permit enriching nitrogen off the column overhead. It helps retain most of the feed CO_2 in the lower portion of the column, leaving the overhead stream enriched in N_2 but largely free of CO_2. As a result, a higher CO_2 tolerance can be allowed in the feed gas. The overhead N_2-enriched stream, usually above 30%, is introduced into the cold box where nitrogen is separated from natural gas. The simple single-column cycle or a double-column cycle is commonly used inside the cold box design. The simple single-column requires a lower investment. However, the double-column cycle is more flexible in responding to a wider variation in the N_2 feed.

Depending upon the concentration of heavy hydrocarbons, a separator or separators are provided in the middle of the chilling step to remove high freezing point components in the feed gas at a proper temperature level. This prevents those components from entering the colder section of the process.

Historically, the rejected N_2 has been vented to the atmosphere with a methane concentration generally specified to be less than 2% to 3% to minimize capital cost. The increasing value of natural gas, however, plus the increasing focus on greenhouse gas emissions, has pressured the industry to reduce methane loses to less than 1%.

It should be noted from figure 10.5 that the methane-N_2 separation step has a narrow operating temperature range compared to the feed gas NGL separation step. Accordingly, temperature control is an important process control variable for the low-temperature or *cold box* portion of the separation process.

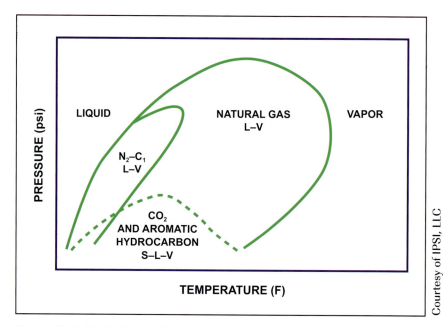

Figure 10.5   Typical natural gas nitrogen phase

REFERENCES

GPSA Engineering Data Book, 12th edition, Gas Processors Suppliers Association, Tulsa, Oklahoma (2004).

Jones, S., Lee, S., Evans, M., and Chen, R., "GPA Research Data Helps Save Time, Money in BG Plant Design," *Oil & Gas Journal* (January 2000).

Kuo, J.C., Elliot, D., Luna-Melo, J., and De Leon Perez, J., "World's Largest N2-generation Plant," *Oil & Gas Journal* (March 2001).

Obrien, J. V., and Maloney, J.J., "Continuous Improvement in Nitrogen Rejection Unit Design," 76th Annual GPA Convention, San Antonio, Texas, March 10–12 (1997).

Trautmann, S., Davis, R., Harris, C., and Ayala, L.,"Cryogenic Technology for Nitrogen Rejection from Variable Content Natural Gas." XIV Convencion Internacional de Gas, Caracas, Venezuela (March 2000).

U.S. Department of Energy, "Nitrogen Removal from Natural Gas," Membrane Technology and Research, Research Report (1999).

All images are copyrighted and may not be reprinted, reproduced, or used in any way without the express written permission of the owner.

Appendix
Figure and Table Credits

Figure		Owner	Web site
1.1	Fluid molecules can be compared to marbles in a glass jar.	PETEX	www.utexas.edu/ce/petex
1.2	Comparison of different temperature measurement scales	PETEX	www.utexas.edu/ce/petex
1.3	Molecules escape and return to the liquid phase in a closed vessel.	PETEX	www.utexas.edu/ce/petex
1.4	Vapor pressure depends on the temperature of the liquid in a closed vessel.	PETEX	www.utexas.edu/ce/petex
1.5	Vapor pressure compared to the boiling temperature of liquid in a closed vessel	PETEX	www.utexas.edu/ce/petex
1.6	Hydrate plug in pipe	Copyright© Petrobras. Photograph by Dr. Alex Freitas	www.petrobras.com
1.7	Vapor pressures of various hydrocarbons	Source: *GPSA Engineering Data Book*, 12th edition	http://gpsa.gasprocessors.com
1.8	Principles of sensible and latent heat	PETEX	www.utexas.edu/ce/petex
1.9	Light molecules vaporize and heavy molecules concentrate in a liquid.	PETEX	www.utexas.edu/ce/petex
1.10	Flow diagram symbols	Source: *GPSA Engineering Data Book*, 12th edition	http://gpsa.gasprocessors.com
1.11	Simple separation system for light and heavy fluid components	PETEX	www.utexas.edu/ce/petex
1.12	Simple separation system with reboiler	PETEX	www.utexas.edu/ce/petex
1.13	Simple separation system with reboiler and condenser	PETEX	www.utexas.edu/ce/petex
1.14	A series separation system	PETEX	www.utexas.edu/ce/petex
1.15	Bubble-cap tray for separation tower	PETEX	www.utexas.edu/ce/petex

Figure	Owner	Web site
1.16 Bubble-cap trays in tower	PETEX	www.utexas.edu/ce/petex
1.17 A tower separation system diagram	PETEX	www.utexas.edu/ce/petex
2.1 Gas processing plant	Copyright© Chevron. All rights reserved.	www.chevron.com
2.2 The functional units inside a gas processing plant	J.C. Kuo	
2.3 A typical feed gas receiving system	J.C. Kuo	
2.4 Diagram of a typical condensate stabilization system	J.C. Kuo	
2.5 Various types of pigs used for pipeline cleaning	Copyright© Girard Industries. All rights reserved.	www.girardind.com
2.6 A typical pig receiver system	J.C. Kuo	
2.7 Pig launcher/receiver	Copyright© Piping Technology & Products, Inc. All rights reserved.	www.pipingtech.com
2.8 Finger-type slug catcher	Copyright© StatoilHydro. Photo by Torstein Tyldum. All rights reserved.	www.statoilhydro.com
2.9 Slug catcher being transported	Copyright© StatoilHydro. Photo by Torstein Tyldum. All rights reserved.	www.statoilhydro.com
2.10 Vertical slug catcher vessel	Copyright© Taylor Forge Engineered Systems. All rights reserved.	www.tfes.com
2.11 Horizontal slug catcher vessel	Copyright© Taylor Forge Engineered Systems. All rights reserved.	www.tfes.com
2.12 Pipe-fitting type slug catcher	Copyright© Taylor Forge Engineered Systems. All rights reserved.	www.tfes.com
3.1 Diagram of a silica gel process	PETEX	www.utexas.edu/ce/petex
3.2 Glycol and propane refrigeration process	J.C. Kuo	
3.3 Glycol/J-T valve cooling process	J.C. Kuo	
3.4 A typical plant refrigeration system	PETEX	www.utexas.edu/ce/petex
3.5 Two different layouts of simple-staged separation process	PETEX	www.utexas.edu/ce/petex
3.6 Diagram of a refrigeration system using staged separation	PETEX	www.utexas.edu/ce/petex
3.7 Gas turbine-driven propane refrigeration compressor in a natural gas plant	Copyright© Elliott Company. All rights reserved.	www.elliott-turbo.com/new
3.8 Chiller diagram	PETEX	www.utexas.edu/ce/petex
3.9 Graph used to determine amount of ethane in propane	PETEX	www.utexas.edu/ce/petex
3.10 Graph used to determine amount of butane in propane	PETEX	www.utexas.edu/ce/petex
3.11 A two-stage refrigeration system	Source: *GPSA Engineering Data Book*, 12th edition	http://gpsa.gasprocessors.com
3.12 A three-stage refrigeration system	Source: *GPSA Engineering Data Book*, 12th edition	http://gpsa.gasprocessors.com

Figure		Owner	Web site
3.13	Diagram of a cascade refrigeration system with ethane and propane	Source: *GPSA Engineering Data Book*, 12th edition	http://gpsa.gasprocessors.com
3.14	Two-stage propane refrigeration system	Source: *GPSA Engineering Data Book*, 12th edition	http://gpsa.gasprocessors.com
4.1	Diagram of a typical aqueous amine treating plant	Source: *GPSA Engineering Data Book*, 12th edition	http://gpsa.gasprocessors.com
4.2	Schematic diagram of the Shell-Paques™ process	Copyright© Shell Global Solutions. All rights reserved.	www.shell.com/globalsolutions
4.3	XTO Shell-Paques™ gas treating plant in Texas	Copyright© Shell Global Solutions. All rights reserved.	www.shell.com/globalsolutions
4.4	Diagram of a typical physical solvent process	Source: *GPSA Engineering Data Book*, 11th edition	http://gpsa.gasprocessors.com
4.5	An integrated natural gas desulfurization plant	Source: *GPSA Engineering Data Book*, 12th edition	http://gpsa.gasprocessors.com
4.6	UOP's spiral wound membrane element	Copyright© UOP, a Honeywell Company. All rights reserved.	www.uop.com
4.7	Element being inserted into the casing	Copyright© UOP, a Honeywell. Company. All rights reserved.	www.uop.com
4.8	A hollow fiber membrane element	PETEX	www.utexas.edu/ce/petex
4.9	A membrane skid for the removal of CO_2 from natural gas	Copyright© UOP, a Honeywell. Company. All rights reserved.	www.uop.com
5.1	Species of elemental S in equilibrium at different temperatures	Source: *GPSA Engineering Data Book*, 11th edition	http://gpsa.gasprocessors.com
5.2	Three-stage modified Claus sulfur-recovery unit	Source: *GPSA Engineering Data Book*, 12th edition	http://gpsa.gasprocessors.com
5.3	A small, package-type, two-stage Claus plant that sends tail gas to an incinerator	Source: *GPSA Engineering Data Book*, 12th edition	http://gpsa.gasprocessors.com
5.4	Diagram of the Shell Claus Off-Gas Treating process (SCOT)	Image supplied from Faraday Management Solution's Sulphur Removal Report to the DTI (UK). Courtesy of Jacobs Engineering Ltd	www.faradaymanagement solutions.com
5.5	SCOT process plants	Copyright© Shell Global Solutions. All rights reserved.	www.shell.com/globalsolutions
6.1	Water content of natural gas varies.	Source: *GPSA Engineering Data Book*, 12th edition	http://gpsa.gasprocessors.com
6.2	Pressure-temperature curves for predicting hydrate formation	Source: *GPSA Engineering Data Book*, 12th edition	http://gpsa.gasprocessors.com
6.3	Physical properties of selected glycols and methanol	Source: *GPSA Engineering Data Book*, 12th edition	http://gpsa.gasprocessors.com
6.4	A typical EG injection system	PETEX	www.utexas.edu/ce/petex
6.5	Freezing temperatures of ethylene glycol-water mixtures	PETEX	www.utexas.edu/ce/petex
6.6	Glycol reboiler temperatures	PETEX	www.utexas.edu/ce/petex
6.7	Hydrate depression versus minimum withdrawal concentration of ethylene glycol	Source: *Dow Chemical Company Gas Conditioning Fact Book*	www.dow.com

Figure		Owner	Web site
6.8a	TEG dehydration unit	Source: *GPSA Engineering Data Book*, 11th edition	http://gpsa.gasprocessors.com
6.8b	Glycol regeneration unit designed to regenerate TEG for natural gas dehydration on an offshore oil and gas production platform	Copyright© Petrex Inc. All rights reserved.	www.petrex.com
6.9	Effect of stripping on TEG concentration	Source: *GPSA Engineering Data Book*, 12th edition	http://gpsa.gasprocessors.com
6.10	Glycol regeneration processes	Source: *GPSA Engineering Data Book*, 12th edition	http://gpsa.gasprocessors.com
6.11	Equilibrium water dew points with various concentrations of TEG	Source: *GPSA Engineering Data Book*, 12th edition	http://gpsa.gasprocessors.com
6.12	TEG reboiler temperatures	PETEX	www.utexas.edu/ce/petex
6.13	Typical desiccant properties	Source: *GPSA Engineering Data Book*, 12th edition	http://gpsa.gasprocessors.com
6.14	Dry-bed dehydration unit schematic	Source: *GPSA Engineering Data Book*, 11th edition	http://gpsa.gasprocessors.com
6.15	Diagram of an adsorption tower	PETEX	www.utexas.edu/ce/petex
6.16	Mass transfer zone for water solid bed adsorption scheme	Source: *Oilfield Processing of Petroleum, Vol I: Natural Gas*, Manning & Thompson (PennWell, 1991)	www.pennwellbooks.com
6.17	Horizontal filter separator	Copyright© Burgess Manning, Inc. All rights reserved.	www.burgess-manning.com
6.18	Mercury removal flow diagram	J.C. Kuo	
7.1	Oil absorption plant systems	Source: *GPSA Engineering Data Book*, 12th edition	http://gpsa.gasprocessors.com
7.2	Lean-oil-to-inlet-gas ratio	PETEX www.utexas.edu/ce/petex	
7.3	A typical low-pressure presaturation system using vapors from rich-oil demethanizer (ROD)	PETEX	www.utexas.edu/ce/petex
7.4	Accumulator	Copyright© Uraltechnostroy Corporation, LLC. All rights reserved.	http://www.uralts.ru/eng
7.5	A residue gas scrubber diagram	PETEX	www.utexas.edu/ce/petex
7.6	Differential pressure indicators	PETEX	www.utexas.edu/ce/petex
7.7	Hot rich-oil flash tank used for methane rejection	PETEX	www.utexas.edu/ce/petex
7.8	Diagram of a ROD	PETEX	www.utexas.edu/ce/petex
7.9	Graph of demethanizer bottom temperature versus lean-oil-to-product ratio	PETEX	www.utexas.edu/ce/petex
7.10	Typical bottom temperature adjustments	PETEX	www.utexas.edu/ce/petex
7.11	The ROD reboiler is heated with hot lean oil in the bottom of the still.	PETEX	www.utexas.edu/ce/petex
7.12	Diagram of a dry still	PETEX	www.utexas.edu/ce/petex
7.13	An oil-reclaiming system design	PETEX	www.utexas.edu/ce/petex

Figure	Owner	Web site
7.14 A distillation test graph showing lean-oil quality	PETEX	www.utexas.edu/ce/petex
7.15 Losses due to poor quality of lean-oil initial boiling point	PETEX	www.utexas.edu/ce/petex
8.1 Lean-oil absorption process and cryogenic process	Source: *Fluid Phase Equilibria*, Vol. 116, Elsevier, Inc.	
8.2 Pressure and temperature to recover 60% ethane	Source: *GPSA Engineering Data Book*, 12th edition	http://gpsa.gasprocessors.com
8.3 Pressure-temperature diagram for the turboexpander process	Source: *GPSA Engineering Data Book*, 12th edition	http://gpsa.gasprocessors.com
8.4 Diagram of a plant using a turboexpander process	Source: *GPSA Engineering Data Book*, 12th edition	http://gpsa.gasprocessors.com
8.5 Methane-ethane binary	Source: *Oil & Gas Journal*, "Petroleum in the 21st Century," 1999	www.ogj.com/index.cfm
8.6 Schematic of gas plant processing	Pervaiz Nasir	
8.7 Deethanizer overhead recycle process	Source: *Oil & Gas Journal* "Petroleum in the 21st Century," 1999	www.ogj.com/index.cfm
8.8 Residue gas recycle process	Source: *Oil & Gas Journal* "Petroleum in the 21st Century," 1999	www.ogj.com/index.cfm
8.9 A 3.5 turboexpander-compressor used to process offshore gas from the Gulf of Mexico	Copyright© Mafi-Trench Company, LLC, part of the Atlas Copco Group. All rights reserved.	www.mafi-trench.com
8.10 Efficiency of turboexpansion cooling	Source: *Hydrocarbon Processing Magazine*, 1970	
8.11 A radial-reaction turbine showing nozzle blades	Copyright© Cryostar SAS. All rights reserved.	www.cryostar.com
8.12 Turboexpander	Copyright© Cryostar SAS. All rights reserved.	www.cryostar.com
8.13 Wheel shaft	Copyright© Cryostar SAS. All rights reserved.	www.cryostar.com
8.14 Active magnetic bearings for typical ABM turboexpanders	Copyright© Cryostar SAS. All rights reserved.	www.cryostar.com
8.15 Core of an aluminum plate	Copyright© Stewart Warner South Wind Corp. All rights reserved.	http://www.stewart-warner.com
8.16 Corrugated fin flow patterns	Copyright© Stewart Warner South Wind Corp. All rights reserved.	http://www.stewart-warner.com
8.17 Cores of plate-fin heat exchanger (PFHE)	Copyright© Chart Energy and Chemicals, Inc. All rights reserved.	www.chart-ind.com
8.18 Components of a brazed aluminum heat exchanger	Copyright© Chart Energy and Chemicals, Inc. All rights reserved.	www.chart-ind.com
9.1 Diagram of a fractionation column	Source: *GPSA Engineering Data Book*, 12th edition	http://gpsa.gasprocessors.com
9.2 Flow through vapor passages	Source: *GPSA Engineering Data Book*, 12th edition	http://gpsa.gasprocessors.com
9.3 Various types of random packing	Courtesy of Beihai Kaite Chemical Packing Company Ltd.	www.kaite-chemical.com

Figure		Owner	Web site
9.4	Tower with various packing materials including structured packing	Copyright© Koch-Glitsch LP. All rights reserved.	www.koch-glitsch.com
9.5	Example of a four-column fractionation plant	PETEX	www.utexas.edu/ce/petex
9.6	Example of a fractionation plant used to produce three products	Source: *GPSA Engineering Data Book*, 12th edition	http://gpsa.gasprocessors.com
9.7	Fractionation plant	Copyright© Dominion. All rights reserved.	www.dom.com
9.8	Schematic of a regenerable caustic process	Source: *GPSA Engineering Data Book*, 12th edition	http://gpsa.gasprocessors.com
10.1	Simplified block flow schematic for a nitrogen rejection facility	Source: *GPSA Engineering Data Book*, 12th edition,	http://gpsa.gasprocessors.com
10.2	A single-column distillation unit	PETEX	www.utexas.edu/ce/petex
10.3	Double-column distillation process	PETEX	www.utexas.edu/ce/petex
10.4	Prefractionation scheme	Copyright© IPSI LLC. All rights reserved.	www.ipsi.com/
10.5	Typical natural gas nitrogen phase	Copyright© IPSI LLC. All rights reserved.	www.ipsi.com/

Table		Owner	Web site
1.1	Physical Properties of Hydrocarbons Involved in Gas Processing	Source: *GPSA Engineering Data Book*, 12th edition	http://gpsa.gasprocessors.com
1.2	Physical Properties of Other Compounds Used in Gas Processing	Source: GPA publication 2145[27]	www.gasprocessors.com
1.3	Calculating Mole Percent PETEX	www.utexas.edu/ce/petex	
2.1	Typical Gas Feeds		
4.1	Approximate Guidelines for Several Commercial Gas Processes	Source: *GPSA Engineering Data Book*, 12th edition	http://gpsa.gasprocessors.com
4.2	Physical Properties of Gas Treating Chemicals	Source: *GPSA Engineering Data Book*, 12th edition	http://gpsa.gasprocessors.com
9.1	Product Specifications for a Southern Louisiana Fractionation Plant	Pervaiz Nasir	
10.1	Components with Allowable Concentrations in a Cryogenic N_2 Rejection Process	PETEX	www.utexas.edu/ce/petex

Glossary

A

AA *abbr*: antiagglomerant.

absolute pressure *n*: the total pressure measured from an absolute vacuum. It equals the sum of the gauge pressure and the atmospheric pressure. Expressed in pounds per square inch absolute (psia).

absolute temperature scale *n*: a scale of temperature measurements in which zero degrees is absolute zero. On the Rankine absolute temperature scale, which is based on degrees Fahrenheit, water freezes at 492° and boils at 672°. On the Kelvin absolute scale, which is based on degrees Celsius, water freezes at 273° and boils at 373°.

absolute zero *n*: defined as precisely 0°K on the Kelvin scale, −273.15°C on the Celsius scale, 0°R on the Rankine scale, and −459.67°F on the Fahrenheit scale.

absorber *n*: a tower or column used to remove various products or emissions from a natural gas stream, which provides contact between the gas and absorption oil.

absorber contactor dehydration system *n*: a system of gas dehydration using glycol desiccants. See *contactor, glycol system*.

absorption *n*: the process of transferring hydrocarbon molecules from the gas phase in the inlet gas to the liquid phase or rich oil. Compare *adsorption, distillation*.

absorption oil *n*: a hydrocarbon mixture used to recover and absorb components from processed gas.

acid gas *n*: carbon dioxide and/or hydrogen sulfide either in or removed from gas.

activated carbon *n*: a highly adsorbent powdered or granular carbon made by carbonization and chemical activation and used in purifying by adsorption. Also known as activated charcoal.

adiabatic expansion *n*: the expansion of a gas or liquid stream from a higher pressure to a lower pressure without heat transfer to the gas, liquid, or surrounding material.

adsorb *v*: to attract molecules of a substance to the surface of another solid substance.

adsorption *n*: the process used to remove water vapor from air or natural gas. Liquid hydrocarbons are recovered from natural gas by passing the gas through activated charcoal, silica gel, or other solids, which extract the heavier hydrocarbons. Treatment of the solid removes the adsorbed hydrocarbons, which are collected and recondensed. Compare *absorption*. See *lean-oil absorption*.

alkali salt *n*: salts that contain a hydroxide ion. Rather than being neutral pH as normal salts usually are, they are weak bases.

alkanolamine *n*: one of a group of viscous, water-soluble amino alcohols. Includes DEA, MEA, DIPA, and MDEA.

amalgam *n*: an alloy of mercury with another metal that is solid or liquid at room temperature according to the proportion of mercury present.

amalgam corrosion *n*: a common type of mercury corrosion that occurs when mercury and aluminum interact in the presence of water. See *corrosion*.

amalgamate *v*: to mix or alloy (a metal) with mercury.

ambient *adj*: the immediate surrounding area. Ambient temperature and ambient humidity are atmospheric conditions that exist in an area at the current time.

American Petroleum Institute (API) *n:* a national trade association that represents America's oil and natural gas industry to the public, Congress and the Executive Branch, state governments, and the media. www.api.org

American Society for Testing and Materials (ASTM International) *n*: a voluntary standards-development organization that addresses the need for international technical standardization for materials, products, systems, and services. ASTM International standards play an important role in the information infrastructure that guides design, manufacturing, and trade in a global economy. www.astm.org

American Society of Mechanical Engineers (ASME) *n*: a not-for-profit professional organization that promotes the art, science, and practice of mechanical and multidisciplinary engineering and allied sciences throughout the world. www.asme.org

amine *n, adj*: an organic compound that contains nitrogen as the key atom and is basic. Amines resemble ammonia in which one or more hydrogen atoms are replaced by organic substitutes such as alkyl and aryl groups.

amine system *n*: a group of processes that use aqueous solutions of various amines to remove hydrogen sulfide (H_2S) and carbon dioxide (CO_2) from gases. Includes an absorber unit and a regenerator unit as well as accessory equipment.

antiagglomerant (AA) *n*: a hydrate inhibitor that limits the size of hydrate crystal formation to a fraction of a millimetre.

API *abbr*: American Petroleum Institute.

API gravity *n:* the measure of the density or gravity of liquid petroleum products on the North American continent, derived from relative density in accordance with the following equations:

API gravity at 60°F = 141.5/specific density – 131.5

API gravity is expressed in degrees, a specific gravity of 1.0 being equivalent to 10°API.

API scale *n:* a scale of liquid gravity measurement units called degrees API, devised and adopted by the American Petroleum Institute. Although the scale is different from an ordinary specific gravity scale, it bears a definite relation to it as follows:

$$°API = \frac{140}{G} - 130$$

where *G* is the specific gravity of the petroleum with reference to water, both at 60°F. The API scale has particular advantages: it provides finer graduations between whole number units and lends itself to schemes for correcting to a temperature standard of 60°F.

aqueous *adj:* water or related to water.

aromatics *n pl:* see *aromatic hydrocarbons*.

aromatic hydrocarbons *n pl:* hydrocarbons derived from or containing a benzene ring. Many have an odor. Single-ring aromatic hydrocarbons are the benzene series (benzene, ethylbenzenes, and toluene). Aromatic hydrocarbons also include naphthalene and anthracene.

ASME *abbr:* American Society of Mechanical Engineers.

associated gas *n:* natural gas that overlies and contacts crude oil in a reservoir. Also called associated free gas. Compare *nonassociated gas.*

ASTM *abbr:* American Society for Testing and Materials. Also *ASTM International.*

atmospheric pressure *n:* the pressure exerted by the weight of the atmosphere. At sea level, the pressure is approximately 14.7 pounds per square inch (101.325 kilopascals), often referred to as 1 atmosphere. Also called barometric pressure.

atomic weight *n:* the weight of a given atom.

B

BAHX *abbr:* brazed aluminum heat exchanger.

bed *n:* several layers of absorbing filters.

biological oxidation process *n:* an energy-producing reaction in living bacteria cells involving the transfer of hydrogen atoms or electrons from one molecule to another.

boiling point *n:* the point at which vapor pressure of a liquid becomes equal to the pressure exerted on the liquid by the surrounding atmosphere. The boiling point of water is 212°F or 100°C at atmospheric pressure (14.7 pounds per square inch gauge or 101.325 kilopascals). Same as boiling temperature.

bottoms *n:* liquid that collects in the bottom of a vessel (tower bottoms, tank bottoms) during a fractionating process or in storage.

brazed aluminum heat exchanger (BAHX) *n:* a heat exchanger used in NGL recovery processes. See *heat exchanger.*

British thermal unit (Btu) n: the amount of heat required to increase the temperature of one pound of water by 1°F.

Btu *abbr:* British thermal unit.

bubble-cap tray *n:* a tray used in a contact tower or absorber to contain liquid. Tray perforations are controlled by valves regulating the gas bubbling up through the liquid. It has the lowest turndown of about 10%.

bubble point *n:* 1. the temperature and pressure at which part of a liquid begins to convert to gas. For example, if a certain volume of liquid is held at constant pressure, but its temperature is increased, a point is reached when bubbles of gas begin to form in the liquid. That is the bubble point. Similarly, if a certain volume of liquid is held at a constant temperature but the pressure is reduced, the point at which gas begins to form is the bubble point. 2. the temperature and pressure at which gas, held in solution in crude oil, breaks out of solution as free gas.

butane *n:* either of two isomeric, flammable gaseous alkanes, C_4H_{10}, obtained usually from petroleum or natural gas and used as fuels.

C

carbon dioxide (CO_2) *n:* a chemical compound composed of two oxygen atoms covalently bonded to a single carbon atom.

carbon disulfide (CS₂) *n*: a colorless, volatile, highly flammable liquid used in solvent extraction processes.

catalytic *adj:* processes used to reduce the toxicity of emissions.

caustic *adj:* able to destroy or deteriorate through a chemical action. See *corrosion.*

caustic treating *n*: a process of treating a substance with a caustic solution to remove impurities. See *regenerative process.*

Celsius scale *n*: the metric scale of temperature measurement used universally by scientists. On this scale, 0° represents the freezing point of water and 100° is its boiling point at a barometric pressure of 760 mm. Degrees Celsius are converted to degrees Fahrenheit by using the following equation:

$$°F = \tfrac{9}{5}\,(°C) + 32$$

The Celsius scale was formerly called the Centigrade scale; now, however, the term Celsius is preferred in the International System of Units (SI).

centrifugal compressor *n*: compresses air or gas by means of mechanical rotating vanes or impellers.

chelate *n*: any class of complex compounds consisting of a central metal atom chemically bonded to a large molecule. More stable than nonchelated compounds of comparable composition.

chiller *n*: a machine that removes heat from a liquid via a vapor-compression or absorption refrigeration cycle.

Claus desulfurization process *n*: see *Claus process.*

Claus process *n*: a gas desulfurizing process in which hydrogen sulfide (H₂S) gas from combustion streams is recovered and turned into salable elemental sulfur. The Claus process has become the industry standard since its invention more than 100 years ago.

Claus sulfur plant *n*: see *Claus process.*

Claus tail gas *n*: see *tail gas.*

closed loop *n*: a system of components that uses feedback from its output for comparison with the desired set point; any difference results in automatic correction of the desired output. Used in automatic control of processes and equipment without human intervention. See *set point.*

CO₂ *n*: see *carbon dioxide.*

coalescer *n*: a mechanical vessel with packing on which liquid can collect for gravity separation.

cobalt (II) sulfide (COS) *n*: used as a catalyst for hydrodesulfurization reactions. Found in nature as the mineral sycoporite.

cold box *n*: a cryogenic gas separation unit with a brazed aluminum heat exchanger and separator. It is usually enclosed by a carbon steel box with perlite insulation and nitrogen purge.

combustion *n*: a chemical reaction with oxygen to produce a flame producing heat and light.

compound *n*: a substance formed by the chemical union of two or more elements in definite proportions; the smallest particle of a chemical compound is a molecule.

compressor *n*: a refrigerating pump mechanism that draws a low pressure on the cooling side of a refrigerant cycle and squeezes or compresses the gas into the high-pressure, condensing side of the cycle.

condensate *n*: a light hydrocarbon liquid obtained by condensation of hydrocarbon vapors in natural gas when pressure and temperature are changed in surface separators. It consists of varying proportions of butane, propane, pentane, and heavier fractions with little or no methane or ethane.

condensate stabilization system *n*: a system used to remove the light components dissolved in the hydrocarbon liquid that comes out of the feed gas reception system.

condensate stabilizer *n*: see *condensate stabilizer system*.

condensate stabilizer reboiler *n*: see *reboiler*.

condensation *n*: the process by which vapors are converted into liquids, chiefly accomplished by cooling the vapors, lowering the pressure on the vapors, or both. Condensation is often the cause of the presence of water in fuels.

condenser *n*: a heat exchanger unit used to turn vapor into liquid.

condensing *v*: the act of a vapor becoming a liquid.

contactor *n*: a vessel designed to bring two or more substances into contact. When natural gas is exposed to liquid amine in a contactor vessel, the carbon dioxide and hydrogen sulfide are absorbed by the amine.

contaminant *n*: 1. a material, usually a mud component, that becomes mixed with cement slurry during displacement and affects it adversely. 2. nonhydrocarbon components (such as liquid mercury) found in natural gas that must be removed.

controller *n*: a mechanism used in a chiller to maintain the proper liquid level.

cooler *n*: a vertical heat exchanger.

corrosion *n*: the reaction between a material, usually a metal, and its environment that produces deterioration of the material.

corrosivity *n*: the tendency or degree of corrosion.

COS *n*: see *cobalt (II) sulfide*.

countercurrent operation *n*: a process in which vapor mixtures are repeatedly brought into contact with liquids having nearly the same composition as the vapors.

critical pressure *n*: the pressure needed to condense a vapor at its critical temperature.

critical temperature *n*: the highest temperature at which a substance can be separated into two fluid phases—liquid and vapor. Above the critical temperature, a gas cannot be liquefied by pressure alone.

cryogenic *adj*: the production or use of very low temperatures.

cryogenic absorption process *n*: a process using chilled hydrocarbon oil to absorb methane and separate nitrogen from natural gas.

cryogenic distillation process *n*: a process used to separate nitrogen from natural gas using condensation and distillation at extremely low temperatures.

cryogenic plant *n*: a gas processing plant operating at temperatures below −50°F that is capable of producing natural gas liquids.

cryogenic process *n*: in material gas processing, the liquefaction of gas at extremely low temperatures in the range of –160°F to –180°F. See also *glycol cryogenic process*.

CS₂ *abbr*: see *carbon disulfide*.

D **DEA** *n*: see *diethanolamine*.

debutanizer *n*: the fractionating column in a natural gasoline plant in which butane and lighter, more volatile components are removed from a hydrocarbon mixture.

deethanizer *n*: equipment used to separate and remove ethane by distillation.

DEG *abbr*: diethylene glycol.

degradation product *n*: a contaminant formed during a reaction such as dehydrogenation.

dehydration *n*: the removal of water from a gas or liquid.

demethanizer *n*: a fractionator used to separate methane and more volatile components from hydrocarbon mixtures. See *rich-oil demethanizer (ROD)*.

demethanizing *n*: removing methane in the rejection system that has been retained by the recovery system and cannot be sold as a liquid product.

depentanizer *n*: a fractionating column in a natural gas plant in which pentane and lighter fractions from a hydrocarbon mixture are removed.

depropanizer *n*: a fractionator used to separate propane and more volatile components from a hydrocarbon mixture.

desiccant *n*: a material that removes water, liquid, or moisture. A dehydration agent.

design cases *n pl*: engineering plans for the design of a gas processing plant based on the nature of the gas and the product to be marketed.

desorb *v*: the removal of moisture and heavy hydrocarbons by an adsorbent.

dew point *n*: the temperature and pressure at which a liquid begins to condense out of a gas. For example, if a constant pressure is held on a certain volume of gas but the temperature is reduced, a point is then reached at which droplets of liquid condense out of the gas. That point is the dew point of the gas at that pressure. Similarly, if a constant temperature is maintained on a volume of gas but the pressure is increased, the point at which liquid begins to condense out is the dew point at that temperature. See *dew-point control*.

dew-point control *n*: the prevention of liquid condensation in the pipeline grid under various pressures and temperatures. Hydrocarbon dew-point control methods include refrigerated low-temperature separation, expanders, Joule-Thomson, and silica gel.

DGA® *n*: see *diglycolamine®*.

diethanolamine (DEA) *n*: colorless, water-soluble, deliquescent crystals, or liquid boiling at 217°C; soluble in alcohol and acetone, insoluble in ether and benzene; used as an absorbent of acid gases. $(HOCH_2CH_2)_2NH$

diethylene glycol (DEG) *n*: a glycol used for water dew-point control and refrigeration.

differential *n*: 1. the difference in quantity or degree between two measurements or units. For example, the pressure differential across a choke is the variation between the pressure on one side and that on the other. 2. the value or volume payment accompanying an exchange of oil for oil. The payment serves as compensation for quality, location, or gravity difference between the oils being exchanged.

differential pressure *n*: the pressure existing across an orifice plate in a fluid flow application; pressure is converted into flow units. The difference between two fluid pressures; for example, the difference between the pressure in a reservoir and in a wellbore drilled in the reservoir, or between atmospheric pressure at sea level and at 10,000 feet (3,048 metres). Also called *pressure differential*.

differential-pressure indicator (DPI) *n*: a pressure-measuring device actuated by two or more pressure-sensitive elements that act in opposition to produce an indication of the difference between two pressures.

diglycolamine® (DGA®) *n*: a solvent used for the removal of CO_2 or H_2S from gases.

diisopropanolamine (DIPA) *n*: $(CH_3CHOHCH_2)_2NH$. A white, crystalline solid with a boiling point of 248.7°C used as a chemical intermediate agent for removal of hydrogen sulfide from natural gas.

DIPA *n*: see *diisopropanolamine*.

discharge pressure *n*: pressure generated on the output side of a gas compressor.

distillation *n*: a separation operation that uses vapor and liquid phases at essentially the same temperature and pressure in two or more existing zones in which each zone moves toward equilibrium with a different concentration in each zone.

downcomer *n*: a method of moving liquid from one tray to the one below in a bubble-tray column. A pipe through which a flow travels downward.

downstream *adv, adj*: 1. in the direction of the stream of fluid moving in a line. 2. commonly used to refer to the refining of crude oil, and the selling and distribution of natural gas and products derived from crude oil.

DPI *abbr*: differential-pressure indicator.

drip gas *n*: a form of gasoline that naturally condenses out of natural gas as it comes from the wells and cools in the field-gathering lines.

drive *n*: a mechanism by which force or power is transmitted in a machine. The means or apparatus for transmitting motion or power to a machine or from one machine part to another.

dry gas *n*: 1. natural gas from wells that does not have a significant content of liquid hydrocarbons or water vapor. 2. all liquid removed from gas in a treatment process.

dry-point *n*: the temperature at which the last liquid evaporates from the distillation vessel. The end of the distillation process.

dry still *n*: a still used to keep lean oil free of water in refrigerated plants to avoid freezing or hydrate problems. Compare *wet still*. See *still*.

E economizer *n*: a stage separator that saves liquid used in the refrigeration system, lets more surge tank liquid propane reach the chillers, and saves horsepower.

EG *abbr*: ethylene glycol.

element *n*: a pure substance that resists ordinary attempts at decomposition. It is usually classed as metal or nonmetal and is expressed by a symbol (e.g., C, carbon: H, hydrogen).

elemental sulfur *n*: sulfur in its native or raw state; a pale yellow, odorless, and brittle material.

energy *n*: the capability of a body for doing work. Potential energy is the capability due to the position or state of the body. Kinetic energy is the capability due to the motion of the body.

enhanced oil recovery (EOR) *n*: 1. the introduction of an artificial drive and displacement mechanism into a reservoir to produce oil unrecoverable by primary recovery methods. To restore formation pressure and fluid flow to a substantial portion of a reservoir, fluid or heat is introduced through injection wells located in rock that has fluid communication with production wells. EOR methods include waterflooding, chemical flooding, most types of gas injection, and thermal recovery. 2. the use of an advanced EOR method.

entrained *v*: to draw in or trap liquid drops into a moving fluid and carry them along in the flow.

EOR *abbr*: enhanced oil recovery.

equilibrium *n*: a state of balance between opposing forces or actions that is either static or dynamic.

ethane *n*: a chemical compound with the formula C_2H_6. At standard temperature and pressure, ethane is a colorless, odorless gas that is used as a fuel.

ethanolamine *n*: see *monoethanolamine*.

ethylene glycol (EG) *n*: a glycol used for water dew-point control and refrigeration.

eutectic freezing point *n*: the lowest temperature at which a liquid can exist.

evaporation *n*: conversion of a liquid to a vapor or gas state.

evaporator *n*: a heat exchanger in which the medium being cooled gives up heat to the refrigerant through the exchanger transfer surface. The liquid refrigerant boils into a gas in the process of the heat absorption.

exchanger *n*: see *heat exchanger*.

exothermic *n*: a chemical reaction that releases heat.

expander *n*: a turbine through which a high-pressure gas is expanded to produce work. Used as a source of refrigeration in industrial processes such as the extraction of ethane, natural gas liquids (NGLs, the liquefication of gas, and other low-temperature processes). Also called a turboexpander.

expansion *n*: the tendency of most materials to expand when heated and to contract when cooled.

expansion turbine *n*: equipment that expands gas or vapor through a turbine to convert the energy content of gas or vapor into mechanical work.

Fahrenheit scale *n*: a temperature scale devised by Gabriel Fahrenheit, in which 32°F represents the freezing point and 212°F the boiling point of water at standard sea-level pressure. Fahrenheit degrees can be converted to Celsius degrees by using the following formula: °C = $\frac{5}{9}$ (°F–32).

feed gas *n*: gas from various reservoirs that is fed into a plant for processing. The content of the feed gas is analyzed to design an individual processing plan. Also called raw feed gas.

feed gas receiving system *n*: the first treating unit in which feed gas is separated into component gases, aqueous liquid, and hydrocarbon liquid for further processing. Also called feed gas reception system.

feedstock *n*: the particular source of oilfield fluids.

fines *n pl*: dust and particles produced during the mechanical destruction of the molecular sieve beds due to high gas velocities, gas channeling, and bed fluidization.

fire point *n*: the lowest temperature at which a liquid produces enough vapors to burn continuously.

flash *n*: the sudden vaporization of a liquid caused by a rapid decrease in pressure.

flashing *v*: the act of quickly vaporizing liquids to stabilize them.

flash point *n*: the temperature at which enough vapors are produced to cause only a momentary flash when ignited. Does not develop into a continuous burn.

flash tank *n*: the vessel in which flashing occurs. See *flash, flashing*.

flow *n*: a current or stream of fluid.

flow control valve *n*: a device used to control the flow of gases, vapor, liquids, slurries, pastes, or solid particles.

flow diagram *n*: a graphical illustration using various symbols to represent, describe, or analyze a process. Same as a schematic.

flow rate *n*: 1. the time required for a given quantity of flowable material to flow a measured distance. 2. the weight or volume of flowable material flowing per unit time. Also known as rate of flow.

flue gas *n*: gas that goes into the atmosphere from a furnace or boiler. Combustion exhaust gas. Same as off-gas.

fluid *n*: a substance that flows freely unless restricted or contained by a barrier. Liquids and gases are considered fluids. Fluids assume the shape of the container in which they are placed.

fractionation *n*: the separation of a chemical compound into its components, or fractions, by their various boiling points by heating them to a temperature at which several fractions of the compound will evaporate. See *distillation*.

fractionation column *n*: a cylindrical piece of equipment where the separated components of hydrocarbons are trapped after distillation.

fractionation tower *n*: see *fractionation column*.

free water *n*: water that can be drained from a vessel and, as a result, separated from the gas or liquid.

Fuller's earth *n*: clay or clay-like material used as a filter.

G

gallons per minute (GPM) *n*: an expression of oil-to-gas ratio of the amount of lean oil circulated per million cubic feet of gas through the absorber.

gas *n*: a compressible fluid that completely fills any container in which it is confined. Technically, a gas will not condense when it is compressed and cooled because a gas can exist only above the critical temperature for its particular composition. Below the critical temperature, this form of matter is known as a vapor because liquid can exist and condensation can occur. The terms gas and vapor may be used interchangeably. The latter, however, should be used for those streams in which condensation can occur and originate from, or are in equilibrium with, a liquid phase.

gas cap *n*: a free-gas phase overlying an oil zone and occurring within the same producing formation as the oil. See *associated gas, reservoir*.

gas chiller *n*: a machine that removes heat from a liquid through a vapor-compression or absorption refrigeration cycle.

gas/gas exchanger *n*: a vertical shell and tube heat exchanger with the feed gas entering the tube channel.

gas/gas heat exchanger *n*: see *gas/gas exchanger*.

gas permeation *n*: the transport of gas molecules through a thin polymeric film from a region of high pressure to one of low pressure. Gas permeation is based on the principle that some gases are more soluble in polymeric membranes and pass more readily through them than other gases.

gas processing *n*: making salable products and treating residue gas to required specifications by separating the components of natural gas.

gas processing plant *n*: a plant that processes natural gas for the recovery of natural gas liquids and other substances.

Gas Processors Association (GPA) *n*: an organization that is a forum for the global gas processing and gas liquids industry. http://www.gasprocessors.com.

Gas Processors Suppliers Association (GPSA) *n*: an outgrowth of the Gas Processors Association that is engaged in meeting the supply and service needs of the natural gas and gas processing industry. http://gpsa.gasprocessors.com.

gas production stream *n*: a body of flowing fluid gas.

gas regeneration *n*: see *regeneration gas*.

gas scrubber *n*: equipment used to remove any water, hydrocarbons, treating chemicals, or corrosion inhibitors from entering the contactor.

gas sub-cooled process (GSP) *n*: the split-vapor process that uses a fraction of the vapor from the cold separator as the top reflux to the demethanizer after substantial condensation and sub-cooling.

gas surge tank (GST) *n*: a container connected to the low-pressure side of a refrigerating system that increases gas volume and reduces rate of pressure change. Also called a surge tank.

gas treatment *n*: the removal of impurities such as hydrogen sulfide and carbon dioxide from raw gas or the green gas stream.

gas treatment unit *n:* equipment for the processing of natural gas.

gauge glass *n*: a glass tube or metal housing with a glass window that is connected to a vessel to indicate the level of the liquid contents.

gauge pressure *n*: 1. the amount of pressure exerted on the interior walls of a vessel by the fluid contained in it (as indicated by a pressure gauge). It is expressed in pounds per square inch gauge or in kilopascals. Gauge pressure plus atmospheric pressure equals absolute pressure. 2. gauge pressure measured relative to atmospheric pressure considered as zero.

glycol *n*: 1. a class of compounds in the alcohol family often used in gas processing. 2. the main ingredient in automotive antifreeze.

glycol cryogenic process *n*: a process in which the feed gas water content is lowered by treating it with a circulating stream of glycol to remove moisture to meet the water dew point. The cryogenic cooling process is used to control the gas hydrocarbon dew point.

glycol/propane system *n*: a system in which glycol is used for water dew-point control and the propane refrigeration system is used for hydrocarbon dew-point control.

glycol system *n*: an absorber contactor dehydration system in which glycol is regenerated in the reboiler and circulated back to the absorption tower, where it contacts the natural gas and removes water as well as hazardous air pollutants.

GPA *abbr*: Gas Processors Association.

GPM *abbr*: gallons per minute.

GPSA *abbr*: Gas Processors Suppliers Association.

gravity *n*: 1. the attraction exerted by the earth's mass on objects at its surface. 2. the weight of a body. 3. an expression of fluid weight stated in degrees of API gravity.

gross heating value *n*: the sum of the normal heating value of methane plus the heating value of the other components of the gas.

GSP *abbr*: gas sub-cooled process.

GST *abbr*: gas surge tank.

H$_2$S *n*: see *hydrogen sulfide*.

heat energy *n*: a process in which substances are transformed to a different state of matter.

heat exchanger *n*: a device that transfers heat energy from two physically separated fluids or mediums of different temperatures.

heating value *n*: the amount of heat produced by the complete combustion of a unit of matter. Gross heating value is the amount of heat derived when the water produced during combustion is condensed. Net heating value is the amount of heat derived when the water produced during combustion is not condensed. Same as heat of combustion.

heat input *n*: reboiler temperature.

heat media *n*: a material, either flowing or static, used to transport heat from a primary source such as combustion of fuel to another material.

heavy ends *n pl*: the heavier hydrocarbons in a natural gas stream. The part of the hydrocarbon mixture with the highest boiling point. Compare *light ends*.

heavy-liquid product *n*: liquid that is not vaporized in the reboiler.

high integrity pressure protection system (HIPPS) *n*: the function of an HIPPS is to protect the downstream equipment against overpressure by closing the source. This is done by closing one or more dedicated safety shutoff valves to prevent further pressurization of the pipe located downstream of these valves.

high-liquid-level alarm (HLA) *n*: a system on the residue gas scrubber to alert the operator when lean oil is about to flood the absorber.

HIPPS *abbr*: high integrity pressure protection system.

HLA *abbr*: high-liquid-level-alarm.

horsepower *n*: a specific amount of mechanical energy.

hydrate *n*: an icy, solid mixture of hydrocarbons and water that can form at temperatures as high as 80°F, depending on the pressure.

hydrate plug *n*: see *plug*.

hydrocarbon dew-point control *n*: a method used to prevent liquid condensation in the pipeline grid under various pressures and temperatures. Includes refrigerated low-temperature separation (LTS), expander, Joule-Thomson (JT), and silica gel.

hydrocarbons *n pl*: organic compounds of hydrogen and carbon, whose densities, boiling points, and freezing points increase as their molecular weights increase. Petroleum is a mixture of many different hydrocarbons.

hydrocarbon treating *n*: a process for the removal of H_2S, CO_2, and other sulfur compounds called acid gases. See *sweetening*.

hydrocarbon treating unit *n*: equipment used in the process of removing H_2S, CO_2, and other sulfur compounds called acid gases.

hydrogen sulfide (H_2S) *n*: a colorless gas with a strong offensive odor that can be detected at a level of two parts per billion. It is highly corrosive and causes metals to become brittle.

hydrolyze *v*: to break a bond in a molecule by adding water.

ideal gas law *n*: a hypothetical ideal gas equation developed by Benoît Paul Émile Clapeyron in 1834 stating that all gases occupy an equal volume at the same conditions of temperature and pressure. The ideal gas law can be stated as $Pv = RT$ or $PV = nRT$ where

$$R = \text{a proportionality factor}$$
$$T = \text{absolute temperature}$$
$$v = \text{volume of one mol of gas}$$
$$n = \text{number of mols of gas}$$
$$V = \text{volume of n mols of gas}$$
$$P = \text{absolute pressure.}$$

immiscible *adj*: not capable of mixing together or attaining homogeneity.

inches of mercury *n*: a measure of atmospheric pressure at a given location based on the height (in inches) of a confined column of mercury in a tube.

inches of water *n*: a measure of hydrostatic, or head pressure, based on the height in inches. Liquid pressure in contact with a column of water in a tube raises the column of water.

initial boiling point *n*: the point at which the mixture starts to boil.

inorganic compounds *n pl*: chemical compounds that do not contain carbon as the principal element (except in the form of carbonates, cyanides, and cyanates). Such compounds make up matter that is neither plant nor animal.

in situ *adj*: in the reaction mixture, unstable molecules must be synthesized within the reaction mixture and cannot be isolated.

instrument *n*: a device for measuring, and sometimes recording and controlling, the value of a quantity under observation.

interstage cooler *n*: equipment to cool the gas between the stages of compression, which removes the heat of compression.

Joule-Thomson (J-T) effect *n*: the change in gas temperature that occurs when the gas is expanded at constant internal energy (enthalpy) from high to low pressure. The effect for most gases at normal pressure, except hydrogen and helium, is a cooling of the gas.

Joule-Thomson (J-T) valve system *n*: see *J-T valve*.

J-T valve *n*: Joule-Thomson valve. A detector cooling device in which a gas under high pressure escapes through an expansion valve in the tank; as the escaped gas liquefies, it cools the site of the sensor.

Kelvin temperature scale *n*: a temperature scale with the degree interval of the Celsius scale and the zero point at absolute zero. On the Kelvin scale, water freezes at 273.16° and boils at 373.16°.

KHI *abbr*: kinetic hydrate inhibitor.

kinetic hydrate inhibitor (KHI) *n*: a hydrate preventative that lowers the rate of formation for a limited time.

KO drum *n*: a knockout drum. A horizontal or vertical piece of equipment designed to remove entrained liquid droplets from gas.

latent heat *n*: the heat that changes the phase of a fluid.   **L**

LDHI *abbr*: low-dosage hydrate inhibitor.

lean gas *n*: 1. residue gas remaining after the recovery of natural gas liquids. 2. unprocessed gas containing little or no recoverable natural gas liquids.

lean glycol *n*: glycol that has been boiled and no longer contains water. It can be reused by pumping it back into the absorber.

lean oil *n*: a hydrocarbon with the most molecules available for absorption while at the same time has characteristics that keep vaporization losses at an economical level.

lean-oil absorption *n*: an increased yield process that separates the various saleable products in natural gas.

level control *n*: maintaining level within specified limits. In a tank or vessel, level control can be accomplished with on/off control or with a proportional-plus reset control. On/off control can be achieved with level switches that control either the inflow or outflow of liquid in the tank.

light ends *n pl*: components of a hydrocarbon liquid mixture that are low boiling and easily evaporated. Compare *heavy ends*.

light hydrocarbons *n pl*: hydrocarbons such as methane, ethane, propane, and butane that have a low molecular weight.

liquefied natural gas (LNG) *n*: the light hydrocarbon part of natural gas, primarily methane, that has been liquefied.

liquefied petroleum gas (LPG) *n*: mainly propane and butane in mixture or separately kept in a liquid state under pressure in a vessel. Also called LP-gas.

liquid *n*: a fluid in a condensed or aqueous state.

liquid expansion *n*: the condition in which a liquid is heated and free to expand in all directions. Same as volume expansion.

liquid level *n*: the uppermost surface of liquid in a tank or vessel; liquid level is controlled by level controllers, measured with liquid-level gauges, and viewed with liquid-level indicators.

liquid separator *n*: a vessel used to separate a vapor-liquid mixture.

LNG *abbr*: liquefied natural gas.

low-dosage hydrate inhibitor (LDHI) *n*: chemicals injected into natural gas to prevent hydrate formation.

low-temperature separation (LTS) *n*: a process where natural gas cools down as pressure is reduced; lowering pressure lowers the temperature to –32°C. Water and the higher hydrocarbons and mercury condense, leaving the natural gas.

LPG *abbr*: liquefied petroleum gas.

LTS *abbr*: low-temperature separation.

M **matter** *n*: anything that possesses mass and occupies space.

MDEA *n*: see *methyldiethanolamine*.

MEA *n*: see *monoethanolamine*.

MEG *n*: see *methyl ethylene glycol*.

membrane process *n*: a gas treatment process in which a polymeric film or membrane is used to separate the acid gas components.

membrane skid *n*: a modular membrane separation unit mounted on a portable platform.

mercaptan *n*: a colorless organic liquid, C_2H_5SH, that has a strong rotten-egg odor and is added to odorless fuel and fuel systems as a warning agent. Hydrosulfide.

mercury removal unit (MRU) *n*: equipment used to remove trace quantities of mercury present in the gas feed to the aluminum heat exchangers.

metal sulfide system *n*: a mercury removal system in which a reactive metal bonded to a support structure, such as carbon or alumina, forms a reactive metal sulfide.

metering *n*: instruments or recorders used to monitor the quantity of process variables used in a process, including temperature, flow, level pressure, etc.

methane *n*: a light, gaseous, flammable paraffinic hydrocarbon, CH_4, that has a boiling point of –258°F and is the chief component of natural gas.

methanol *n*: a chemical compound with chemical formula CH_3OH (abbreviated MeOH). It is the simplest alcohol and is a light, volatile, colorless, flammable liquid with a distinctive odor.

methyldiethanolamine (MDEA) *n*: an alkanolamine used in tail-gas treating and hydrogen sulfide enrichment units for selectively removing hydrogen sulfide from gas streams containing carbon dioxide.

methyl ethylene glycol (MEG) *n*: a glycol used for water dew-point control and refrigeration.

million standard cubic feet (MMscf) *n*: a standard measurement in the gas industry based on 60°F and 14.7 psia.

million standard cubic feet per day (MMscf/d) *n*: a measure of gas volume at 60°F and 14.7 psia per day.

miscibility *n:* a property of liquids that is evidenced by their ability to mix.

miscible *adj*: 1. capable of being mixed. 2. capable of mixing in any ratio without separation of the two phases.

mode *n*: one of several alternative conditions or methods of operation of a device.

mole *n*: the fundamental unit of mass of a substance. A mole of any substance is the number of grams or pounds indicated by its molecular weight.

molecular sieve *n*: a material containing tiny pores of a precise and uniform size used as an adsorbent or desiccant for gases and liquids that operates on a molecular level. Also called mol sieve.

molecular weight *n*: the sum of the atomic weights in a molecule. For example, the molecular weight of water, H_2O, is 18 because the atomic weight of each of the hydrogen molecules is 1 and the atomic weight of oxygen is 16.

molecule *n*: the smallest particle of a substance that retains the properties of the substance; it is composed of one or more atoms.

mole percent *n*: the ratio of the number of moles of one substance to the total number of moles in a mixture of substances, multiplied by 100 (to put the number on a percentage basis).

monoethanolamine (MEA) *n*: a colorless liquid, $NH_2(CH_2)_2OH$, used in the purification of petroleum, sweetening natural gas, and the production of ethylene amines. Also called ethanolamine.

MRU *abbr*: mercury removal unit.

mutually soluble *adj*: able to be mixed.

natural gas *n*: a highly compressible, highly expansible mixture of hydrocarbons with a low specific gravity and occurring naturally in a gaseous form. Besides hydrocarbon gases, natural gas might contain appreciable quantities of nitrogen, helium, carbon dioxide, hydrogen sulfide, and water vapor. Although gaseous at normal temperatures and pressure, the gases making up the mixture that is natural gas are variable in form and might be found either as gases or liquids under suitable conditions of temperature and pressure.

natural gas liquids (NGL) *n pl*: those hydrocarbons liquefied at the surface in field facilities or in gas processing plants. Natural gas liquids include propane, butane, and natural gasoline.

natural gas processing plant *n*: see *gas processing plant*.

net heating value *n*: the difference between the gross heating value and the heat energy required to vaporize water trapped in the gas. Also called low heating value.

NGL *abbr*: natural gas liquid.

nitrogen rejection *n*: a process used to remove nitrogen from natural gas to maintain a desired Btu value for marketing. See *British thermal unit*.

nitrogen rejection unit (NRU) *n*: equipment in which nitrogen is removed from feed gas.

nonassociated gas *n*: gas in a reservoir that contains no oil. Compare *associated gas*.

nonregenerative process *n*: a method of treating used when only small amounts of contaminants are removed and/or high purity of treated gas is desired. Compare *regenerative process*.

NRU *abbr*: nitrogen rejection unit.

O **off-gas** *n*: fumes or exhaust gases produced by combustion. See *flue gas*.

oil whip *n*: a condition arising from the latter stages of oil whirl. It is a dangerous condition because the rotor uses up the entire bearing clearance and is in direct metal-to-metal contact that will wear away the bearing rapidly and destroy the rotor if not corrected. See *oil whirl*.

oil whirl *n*: an uneven oil distribution (oil wedge) around the shaft in the journal bearing.

open loop *n*: a control system that does not incorporate automatic feedback but uses manual adjustments to maintain the desired function.

operating factor *n*: the percentage of time equipment performs the function for which it is designed.

overhead product *n*: the top or main product produced by a fractionation column. See *fractionation column*.

oxidation *n*: the chemical interaction of a substance with oxygen (O_2) or an oxygen-containing material that adds oxygen atom(s) to the compound being oxidized.

P **packed column** *n*: a fractionation or absorption column filled with material that comes in contact with a gas to promote separation. Compare *bubble-cap trays*.

packing *n*: small solid shapes over which liquid and vapor flow. Expanded metal or woven mats are also used as packing. Also called structure packing.

paraffin *n*: a saturated aliphatic hydrocarbon having the formula C_nH_{2n+2} (e.g., methane, CH_4, and ethane, C_2H_6). Heavier paraffin hydrocarbons (i.e., $C_{18}H_{38}$ and heavier) form a wax-like substance that is called paraffin.

petroleum *n*: a substance occurring naturally in the earth in solid, liquid, or gaseous state and composed mainly of mixtures of chemical compounds of carbon and hydrogen, with or without other nonmetallic elements such as sulfur, oxygen, and nitrogen.

petroleum geology *n*: the study of oil- and gas-bearing rock formations. It deals with the origin, occurrence, movement, and accumulation of hydrocarbon fuels.

phase *n*: a portion of a physical system that is liquid, gas, or solid, that is homogeneous throughout, has definite boundaries, and can be separated from other phases.

PI *abbr*: pressure indicator.

pig *n*: a device inserted into a pipeline to perform several functions such as cleaning or internal inspection.

pigging *v*: the act of using a mechanical pig to clean, inspect, or remove accumulated liquid in a pipeline. See *pig*.

pig receiver *n*: an arrangement of pipes that allows pigs to be removed from a pipeline without stopping flow. See *pig*.

pig system *n*: a system used to periodically remove the accumulation of liquids at low spots in the pipeline for pipeline cleaning or inspections.

pipeline gas *n*: a gas that meets the minimum specifications of a transmission company.

pipeline grid *n*: a design that provides an integrated natural gas delivery system to the consumer.

plug *n*: any object or device that blocks a hole or passageway, or the movable part of a valve used to block or restrict the passageway of fluids in the valve.

potassium carbonate *n*: a white salt (K_2CO_3) that is basic in solution. Called potash.

pounds per square inch (psi) *n*: a measure of pounds force exerted within one square inch.

pounds per square inch absolute (psia) *n*: the true pressure of a contained fluid that is the sum of pounds per square inch gauge and atmospheric pressure.

pounds per square inch gauge (psig) *n*: a measure of the gauged pressure of a fluid in a containment vessel.

power *n*: 1. the time rate of doing work, watts, horsepower, ft-lbs/second. 2. the value assigned to a mathematical expression and its exponent.

ppmv *n*: parts per million by volume.

prefractionator *n*: a fractionator that separates the lightest hydrocarbon components.

presaturation *n*: a method that removes the heat of absorption and minimizes the temperature in the absorber by taking low-pressure vapors from the demethanizer.

pressure *n*: the stress or force exerted uniformly in all directions; it is usually measured in terms of force exerted per unit area, such as pounds per square inch (psi) or newtons per square metre.

pressure controller *n*: an electronic or pneumatic device that maintains a constant pressure at a specific point in a process, such as a pressure-operated valve.

pressure differential *n*: see *differential pressure*.

pressure indicator (PI) *n*: a gauge that measures and monitors pressure.

pressure letdown *n*: depressurization.

pressure regulator *n*: a device for maintaining pressure in a line, downstream from the device.

pressure swing adsorption (PSA) *n*: a technology used to separate nitrogen under pressure according to its molecular characteristics and attraction to an adsorbent material at near-ambient temperatures.

process engineer *n*: an engineer who develops economical industrial processes.

process variables *n pl*: quantities such as pressure, temperature, flow, and level in a process system.

product *n*: the various components produced in the gas processing plant. Refers generally to the desirable components.

production *n*: 1. the phase of the petroleum industry that deals with bringing the well fluids to the surface and separating them, and with storing, gauging, and otherwise preparing the product for the pipeline. 2. the amount of oil or gas produced in a given period.

propane *n*: a paraffinic hydrocarbon (C_3H_8) that is a gas at ordinary atmospheric conditions but is easily liquefied under pressure. It is a constituent of liquefied petroleum gas.

PSA *abbr*: pressure swing adsorption.

Q   **quench column** *n*: a tower in which off-gases from the reactor are cooled.

R   **range** *n*: the calibration of process measuring instruments between its lower value (zero) and upper value (span).

Rankine temperature scale *n*: a temperature scale with the zero point at absolute zero. On the Rankine scale, water freezes at 491.60°R and boils at 671.69°R.

raw feed gas *n*: see *feed gas*.

reaction kinetics *n*: the rates and mechanisms of chemical reactions for a large number of gas-phase chemical reactions.

reagent *n*: a substance or compound consumed during a chemical reaction.

reboiler *n*: a heating unit for a fractionating tower that supplies additional heat to the lower portion of the tower.

reciprocating compressor *n*: equipment that uses pistons driven by a crankshaft to deliver gases at high pressure.

reclaimer *n*: a filtration system used to remove corrosive degradation products caused by the reaction of MEA with COS and CS_2.

recompression *n*: the process of compressing sales gas to pipeline pressure for delivery.

reconcentrator *n*: equipment used to evaporate moisture collected from feed gas and restore the original concentration of glycol.

recovery *n*: the first system in an oil absorption plant in which products are removed from inlet gas by a combination of refrigeration and oil absorption. Some unwanted methane remains. See *rejection, separation*.

rectification section *n*: the top section of the separation tower. Also called rectifying section.

recycle *v*: returning part of a gas processing stream to a point upstream to enhance recovery or control.

redox *n*: describes all chemical reactions in which atoms have their oxidation state changed.

redox process *n*: any chemical reaction that involves oxidation and reduction. In nonamine processes, H_2S is directly converted to elemental sulfur and water by reacting with oxygen.

reflux *n*: in a distillation process, that part of the condensed overhead stream that is returned to the fractionating column as a source of cooling. *v*: in distillation, extraction of fluids from a core, to use a solvent to flow over a core sample a second time to clean it.

refrigeration system *n*: a system that operates by continuous circulation of a refrigerant in a four-step cycle: expansion, evaporation, compression, and condensation.

regeneration gas *n*: gas produced by reacting natural gas with a carrier gas (such as steam) over a steam reforming catalyst.

regenerative molecular sieve system *n*: a mercury removal system using silver deposited on the molecular sieve pellets. The silver reacts with the mercury producing an amalgam that is insoluble in hydrocarbon liquid.

regenerative process *n*: treating used to remove large quantities of contaminants by chemical reaction (amine-based solvents or nonamine-based solvents), physical absorption, mixed chemical/physical absorption, and adsorption on a solid.

Reid vapor pressure (RVP) *n*: a measure of vapor pressure for petroleum fractions and their blends, measured in pounds per square inch (psi). The higher the RVP, the more volatile the product. The test procedure is reported as ASTM test method D 323-89 in the 1990 Annual Book of ASTM Standards, Section 5, Petroleum Products, Lubricants, and Fossil Fuels.

rejection *n*: after recovery, the second system in an oil absorption plant in which the remaining methane is eliminated from the product stream, leaving only the desired product and absorption oil. See *recovery, separation*.

reservoir *n*: a subsurface, porous, permeable rock body in which oil and/or gas have accumulated.

resonance *n*: the vibration in mechanical parts caused by high speeds or oil film.

retrograde condensation *n*: condensation or vaporization that is the reverse of expected gas performance. A decrease in pressure or increase in temperature causes condensation. An increase in pressure or decrease in temperature causes vaporization.

rich glycol *n*: a glycol and water mixture used to prevent hydrate formation.

rich oil *n*: lean oil that has absorbed heavier hydrocarbons from natural gas.

rich-oil demethanizer bottoms *n*: the bottom section of the separation tower between the feed gas and the reboiler where light gas components are removed. Also called stripping section.

rich-oil demethanizer (ROD) *n*: part of the rejection system. A vessel in which rich-oil methane is rejected while retaining most of the ethane.

rich-oil fractionator *n*: see *still.*

ROD *abbr*: rich-oil demethanizer.

royalty *n*: the portion of oil, gas, and minerals retained by the lessor upon execution of a lease; or their cash value paid by the lessee to the lessor or to one who has acquired possession of the royalty rights, based on a certain percentage of the gross production from the property free and clear of all costs, except taxes.

RVP *abbr*: Reid vapor pressure.

S

safety instrumented systems (SIS) *n*: instrumented systems to protect against overpressure; failure of the systems can result in the release of hazardous chemicals and/or cause unsafe working conditions.

schematic *n*: see *flow diagram.*

SCOT *abbr*: Shell Claus Off-Gas Treating process.

scrubber *n*: see *gas scrubber.*

sensible heat *n*: the heat that changes the temperature of a fluid without changing its phase to gas or solid.

separation *n*: after recovery and rejection, the third system in oil-absorption plants that is used to separate gas products. See *recovery, rejection. v*: processing various marketable products from natural gas.

separation tower *n*: see *tower with trays.*

set point *n*: the desired value or output in a control system. Establishes the reference in a closed-loop control system.

shell-and-tube exchanger *n*: a type of heat exchanger consisting of tubes in a container called a shell. Transfers heat from a fluid on one side of a barrier to a fluid on the other side, without bringing the fluids into direct contact.

Shell Claus Off-Gas Treating (SCOT) process *n*: a process for improving the efficiency of a Sulfur Recovery Unit (SRU; Claus Unit). Sulfur recovery of the SRU can reach about 95%–99.9%.

Shell-Paques process *n*: a biodesulfurization process. A gas stream containing H_2S contacts an aqueous soda solution containing thiobacillus bacteria in an absorber. The soda absorbs the H_2S and is transferred to an aerated atmospheric tank where the bacteria biologically converts the H_2S to elemental sulfur. This process is ideally suited to environmentally sensitive areas where venting, incineration, or reinjection of H_2S are not desirable options. See *thiobacillus bacteria.*

Shell Twister™ Supersonic separator *n*: see *Twister™ Supersonic separator.*

sieve tray *n*: a tray installed in an absorber tower or fractionating column that is similar to a bubble-cap tray, except that it has only holes not bubble caps through it. This type of tray is more efficient than the bubble-cap tray or the valve tray and less expensive than either; however, it does not operate properly over a wide range of flow rates. See *molecular sieve.*

silica gel *n*: a granular desiccant used to remove water from gas. A dehydration method.

silica gel process *n*: a dehydration process used for water as well as hydrocarbon dew-point control in which water and heavy hydrocarbons are removed from feed gas on multiple beds of silica gel adsorbent.

SIS *abbr*: safety instrumented systems.

slip *v*: to reject a substance such as CO_2 in the absorption process.

slug *n*: an amount of liquid accumulation in a pipeline.

slug catcher *n*: an arrangement of gas plant piping made to intercept the slug of liquid and separate it out of the gas stream.

solubility *n*: the degree to which a substance will dissolve in a particular solvent.

solution reclaimer *n*: see *reclaimer*.

sour gas *n*: natural gas having more than 1½ grains of hydrogen sulfide per 100 cubic feet or more than 30 grains of total sulfur per 100 cubic feet. It is unfit for domestic use without processing.

specific gravity *n*: the ratio of the mass of a given volume of substance to the mass of an equal volume of water. It is a dimensionless ratio that provides the relative weight of a liquid to that of water, such as oil.

specific viscosity *n*: the ratio of the absolute viscosity of a substance to that of a standard fluid, such as water, with the viscosity of both fluids being measured at the same temperature.

SRU *abbr*: sulfur recovery unit.

stabilization column *n*: units used to remove the light components dissolved in the hydrocarbon liquid that come out of the feed gas reception system. See *condensate stabilization system*.

stabilized condensate *n*: condensate stabilized at a specific vapor pressure in a fractionation system.

stabilizer *n*: a fractionation column designed to reduce the vapor pressure of a liquid stream.

staged separation *n*: a process used to increase oil recovery in the stock tank by producing the well through several separators in a series.

standard cubic foot (scf) *n*: a measure of a quantity of gas equal to a cubic foot of volume at 60°F and either 14.696 pounds per square inch (1 atm) of pressure.

sterically hindered *adj*: see *steric hindrance*.

steric hindrance *n*: the prevention or retardation of a chemical reaction caused by the arrangement of atoms in a molecule. It can be useful to change the reactivity pattern of a molecule by stopping unwanted side reactions.

steric resistance *n*: see *steric hindrance*.

still *n*: a product-condensing facility or vessel in which hydrocarbon distillation is done. Also called a rich-oil fractionator.

stream *n*: an unbroken flow of gas or particle matter.

strip *v*: the process of removing light components in the gas through vaporization.

stripping gas *n*: a dry, low-pressure gas used to remove the final amounts of water in a dehydration process.

stripping section *n*: see *rich-oil demethanizer bottoms*.

structure packing *n*: see *packing*.

sulfide *n*: any of the three classes of chemical compounds that contain the element sulfur. A compound of sulfur that contains the S^{-2} ion.

sulfur pits *n*: subsurface containers used to hold liquid sulfur streams.

sulfur recovery unit (SRU) *n*: equipment used to control tail-gas emissions. Acid gas is fed into the SRU from the treating plant where H_2S and other sulfur compounds are converted and recovered as nontoxic elemental sulfur (S).

surge tank *n*: see *gas surge tank*.

sweetening *n*: see *hydrocarbon treating*.

sweet gas *n*: raw natural gas with low concentration of sour compounds, such as hydrogen sulfide and carbon dioxide.

T

tail gas *n*: gas leaving a plant. Quantities of sulfur compounds that must be reduced before the dispersion of the gases to the atmosphere. The sulfur compounds in the tail gases are principally unreacted hydrogen sulfide (H_2S) and sulfur dioxide with traces of carbonyl sulfide and carbon disulphide. See *Claus process*.

TEG *abbr*: triethylene glycol.

temperature *n*: a measure of heat or the absence of heat, expressed in degrees Fahrenheit or Celsius. The latter is the standard used in countries on the metric system.

temperature indicator (TI) *n*: a gauge that measures and monitors temperature.

tertiary amine *adj*: an amine created when all of the hydrogens in an ammonia molecule have been replaced by hydrocarbon groups.

tetraethylene glycol (TREG) *n*: a glycol used for water dew-point control and refrigeration.

thermocouple *n*: a device consisting of two dissimilar metals bonded together, with electrical connections to each. When the device is exposed to heat, an electrical current is generated, the magnitude of which varies with the temperature. It is used to measure temperatures higher than those that can be measured by an ordinary thermometer, such as those in engine exhausts.

thermodynamic inhibitor *n*: a chemical injected into natural gas that lowers the temperature of hydrate formation.

thiobacillus bacteria *n*: small rod-shaped bacteria living in oxidizing sulfur created from hydrogen sulfide accumulations.

TI *abbr*: temperature indicator.

TIC *abbr*: total installation cost.

total installation cost (TIC) *n*: the cost of installing something (as equipment).

tower *n*: see *quench column*.

tower with trays *n*: a gas separation system that reduces the amount of required equipment. Various types include bubble cap, valve, sieve, and perforated. See *separation tower, valve trays*.

train *n*: a series of units functioning together to complete a complex process.

trays *n*: see *tower with trays, valve trays*.

treating *n*: refers to any of several processes to remove impurities from natural gas.

treating plant *n*: a facility designed to remove undesirable impurities from natural gas to render the gas marketable.

TREG *abbr*: tetraethylene glycol.

triethylene glycol (TEG) *n*: a glycol used for water dew-point control and refrigeration.

turboexpander *n*: a turbine through which a high-pressure gas is expanded to produce work. Also called an expander.

turboexpansion *n*: refrigeration in industrial processes such as the extraction of ethane, natural gas liquids, the liquefaction of gases, and other low-temperature processes.

turndown ratio *n*: the highest and lowest effective capacity of the processing system. It is calculated by dividing the maximum output by the minimum output at which controlled and efficient combustion can be maintained. A 4:1 turndown means the system is running at one-quarter of maximum capacity.

Twister™ Supersonic separator *n*: treats produced gas at supersonic velocities, extracting water and hydrocarbon liquids. The process requires no chemicals and has no moving parts. Developed by Shell.

U

upstream *n*: the point in a line or system situated opposite the direction of flow. *adv.* in the direction opposite flow in a line.

V

vacuum *n*: 1. a space that is theoretically devoid of all matter and that exerts zero pressure. 2. a condition that exists in a system when pressure is reduced below atmospheric pressure.

valve *n*: a device used to control the rate of flow in a line to open or shut off a line completely, or to serve as an automatic or semiautomatic safety device. Those used extensively include the check valve, gate valve, globe valve, needle valve, plug valve, and pressure-relief valve.

valve tray *n*: a tray used in a contact tower or absorber to contain liquid. Tray perforations are controlled by valves regulating the gas bubbling up through the liquid. It has the lowest turndown of about 35%. See *bubble-cap trays, tower with trays*.

vapor *n*: a substance in a gaseous state that can be liquefied by compression or cooling. The terms vapor and gas are technically interchangeable.

vaporization *n*: 1. the act or process of converting a substance into the vapor phase. 2. the state of substances in the vapor phase.

vaporizing *v*: the act of a liquid turning into a vapor.

vapor pressure *n*: the pressure exerted by the vapor of a substance when the substance and its vapor are in equilibrium. Equilibrium is established when the rate of evaporation of a substance is equal to the rate of condensation of its vapor.

viscosity *n*: a measure of the resistance of a fluid to flow. Resistance is brought about by the internal friction resulting from the combined effects of cohesion and adhesion. The viscosity of petroleum products is commonly expressed in terms of the time required for a specific volume of the liquid to flow through an orifice of a specific size at a given temperature.

viscous *adj*: having a high resistance to flow.

volatility *adj*: how easily a given fluid will vaporize.

volume *n*: the amount of a substance that occupies a particular space.

volume expansion *n*: the condition in which a liquid is heated and free to expand in all directions. Same as liquid expansion.

volume percent *n*: same as mole percent.

W

water dew-point control *n*: the prevention of liquid condensation in the pipeline grid under various pressures and temperatures. Water dew-point control methods include the silica gel, glycol, and molecular sieve.

weir *n*: a barrier or enclosure that maintains a measured flow of liquid or vapor.

wet still *n*: a still that uses steam vapor to act as a stripping medium. See *still*. Compare *dry still*.

Wobbe number *n*: a number proportional to the heat input into a burner at constant pressure.

working fluid *n*: refrigerant coolant material.

Index

absolute pressure, 3

absolute zero, 2

absorbers, 98

absorber workings, 99–100.

absorption, 36, 51

absorption oil, 99

activated carbon process, 62

adiabatic expansion (JT), 148

adsorbent life, 89

adsorption on a solid, 61–62

air-to-fuel ratio, 13

alkali salt solutions, 51

alkanolamines, 51, 141

amalgamation, 92

amalgam corrosion, 92

ambient temperature, 58

American Petroleum Institute (API), 28

American Society for Testing and Materials (ASTM International), 109

American Society of Mechanical Engineers (ASME), 28

antiagglomerants (AAs), 76

API gravity, 3

applications, gas processing general, 13–21

applications, NGL recovery-cryogenic
ethane-recovery process, 121–122
propane-recovery process, 120
turboexpander process, 115–119

aromatics, 26

associated gas, 1

biological oxidation process, 57

boiling point, 6

bottoms temperature, 106

British thermal units (BTUs), 10

bubble cap, 18

bubble-point line, 116

bubble-points, 131

butane splitter column, 137

carbon disulfide (CS_2), 53

case study, 95

catalytic recovery, 68, 70

catalytic step, 67

caustic treating, 141

centrifugal compressors, 32

chelate solution, 51, 57

chemical reactions
amine-based solvents, 52–53

biochemical processes, 57–58
diethanolamine (DEA), 53
diglycolamine (DGA), 53
formulated amines, 54
hindered amines, 54
hot potassium carbonate process, 57
methyldiethanolamine (MDEA), 54
monoethanolamine (MEA), 53
nonamine-based processes, 57–58
redox process, 57
chillers, 35, 44
chilling, 148
Claus, Carl Friedrich, 67
Claus off-gas treating, 70
Claus process, 54
closed loop propane refrigeration system, 38
coalescent liquid level problem, 95
coalescers, 141
cobalt (II) carbonyl sulfide, 53
cold box, 152
combustion, 12–13
comparison of dew-point processes, 40
composition, 10
compressors, 1, 32
condensate product storage tanks, 28
condensate stabilizer reboilers, 32
condensate stabilizers, 32
condensers, 16
condensing, 5
contactor vessel, 52
contaminants, 1
controllers, 44
coolers, 38
core exchangers, 127
corrosion, 25, 65
corrosivity, 53
cost estimate, 35
countercurrent operation, 131
critical pressures, 3
critical temperatures, 3
cryogenic absorption, 145
cryogenic distillation, 146, 148
cryogenic NRU processes
    chilling, 148
    cryogenic distillation, 148
    nitrogen rejection units (NRUs), 146–148
    pretreatment, 147–148
    recompression, 148
cryogenics, 127–129

debutanizer, 28

debutanizer (DeC$_4$) column, 137
deethanizer, 120
deethanizer (DeC$_2$) column, 137
degradation products, 53
dehydration, 4, 73–80
dehydration and mercury removal, 73–95
dehydration methods
    design basis and specifications, 92–93
    design issues, 88
    liquid desiccants, 80, 82–84
    mercury removal unit (MRU), 90
    molecular sieve process, 85–86
    silica gels and activated ammonia, 85
    solid desiccants, 84–85
deisobutanizer (DIB) column, 137
demethanizing, 104
depentanizer, 28
depropanizer (DeC$_3$) column, 137
desiccants, 147
design basis and specifications for dehydration
  and mercury removal
    about generally, 92
    case study, 95
    design considerations, 94–95
    equipment selection and designs, 95
    gas dehydration units, 92
    general considerations, 92
    technology selection, 92
design basis and specifications for treatment units
    condensate product storage tanks, 28
    feed case basis, 26
    feed gas, 26–28
    product specifications, 27
design cases, 26
design considerations
    design basis and specifications for dehydration
      and mercury removal, 94–95
    free-liquid removal, 94
    mercury removal costs, 94
    MRU location, 94
    pressure drop, 94
    superficial velocity and residence time, 95
design issues
    adsorbent life, 89
    dehydration methods, 88
    poor outlet water dew point, 88–89
    pressure drop, 89
    switching valves for adsorption and regeneration
      operation, 90
desorption, 37
dew-point control, 25
dew-point control and refrigeration systems, 35–49
    chillers, 44

cost estimate of, 35
   economizers, 42
   multiple stage refrigeration, 46
   problems, 44–46
   process descriptions, 35–46
   refrigeration systems, 41
dew-point depression, 35
diethanolamine (DEA), 52
diethylene glycol (DEG), 37
differential pressure indicator (DPI), 102
diisopropanolamine (DIPA), 60
distillation, 108
downcomers, 131
downstream processing, 23
drip gas, 97
drive assembly, 45
dry gas loss, 12
dry point temperature, 109
dry stills, 108

economizers, 42
elemental sulfur, 57
entrained water vapor, 12
equilibrium, 5
equipment selection and design
   condensate stabilizer reboilers, 32
   condensate stabilizers, 32
   feed gas, 28
   gas and liquid heaters, 32–33
   mercury adsorber, 95
   mercury removal after-filter, 95
   pig receivers, 28
slug catchers, 30
stabilizer overhead compressors, 32
ethane-recovery process, 121–122
ethylene glycol (EG), 37, 77
eutectic freezing point, 77
evaporation, 41
evaporators, 35
exchangers, 127
exothermic process, 67
expanders, 35

Fahrenheit scale, 2
feed gas basis, 26
feed gas receiving and condensate stabilization, 23–33
   design basis and specifications for treatment
     units, 26–28
   equipment selection and design, 28
   treating and processing, 23–26

feedstock, 1
filtration, 65
fire point, 13
flammability, 13
flashing, 32
flash point, 13
flash tank, 77
flooding, 140
flow-control valve, 25
flow diagrams, 14, 16, 18, 19-20
flow rate, 28
flue gas, 12
fluid, 1
fluid properties, 1–4
   gravity, 3
   miscibility, 3
   pressure, 3
   solubility, 4
   temperature, 2–3
foaming, 65
fractionation, 131–133
fractionation and liquid treating, 131–143
fractionation, 131–133
   monitoring of fractionalization plants, 139–140
   NGL fractionation plants, 134–136
   NGL product treating, 141–143
   operating problems, 140
   packed columns, 134
   product specification, 138
fractionation column, 26
free-liquid removal, 94
freezing, 103
freezing point, 6
Fuller's earth, 143
fundamentals
   combustion, 13
   composition, 10
   fluid properties, 1–4
   heat energy, 10–12
   hydrates, 7–9
   ideal gas law, 4–6

gallons per minute (gpm), 100
gas and liquid heaters, 32–33
gas cap, 1
gas chiller, 41
gas dehydration units, 92
gas/gas exchanger, 38
gas permeation, 62
Gas Processors Suppliers Association (GPSA), 6
gas production streams, 1

gas sub-cooled process (GSP), 119

gas surge tank, 41

gas treatment, 1

gas treatment unit, 26

gauge pressure, 3

general considerations for dehydration and mercury removal, 92

general operating considerations for gas treating, 65

glycol, 4

glycol cryogenic process, 38

glycol injection problems, 78–80

glycol/J-T valve cooling process, 38–40

glycol/propane system, 37–38

GPSA Engineering Data Book, 28, 84

gravity, 3

gross heating value, 12

heat energy, 10–12

heating value, 12

heavy-liquid product, 20

high-integrity pressure protection system (HIPPS), 28

high-liquid level alarm (HLA), 102

horsepower, 10

hot-rich oil flash tank, 104

hydrate plug, 7

hydrates, 7, 8–9, 73

hydrocarbon dew-point control, 35

hydrocarbon fluids, 1

hydrocarbon treating, 51–65

    adsorption on a solid, 61–62

    chemical reactions, 51–58

    gas treating processes, 51

    general operating considerations for gas treating, 65

    membrane processes, 62–64

    mixed chemical/physical absorption processes, 60

    physical absorption processes, 58–60

hydrocarbon treating units, 27

hydrogen sulfide, 1

ideal gas law

    about generally, 4

    boiling point, 6

    freezing point, 6

    liquid phase, 5

    vapor pressure, 5–6

immiscible liquids, 3

incorrect coalescer liquid level, 95

inhibitor injection, 76–80

initial boiling point, 13

inlet separation, 65

interstage coolers, 32

Joule-Thomson (J-T), 35

Joule-Thomson (J-T) expansion, 117

Joule-Thomson (J-T) valve system, 38

Kelvin scale, 2

kinetic hydrate inhibitors (KHIs), 76

KO drum (knockout drum), 36

latent heat, 11

lean glycol, 77

lean oil, 44

lean oil absorption

    about generally, 97

    absorber workings, 99–100.

    potential problems with, 102–103

    presaturation, 100

    recovery system, 98–103

level controllers, 44

level control valve, 38

liquid desiccants, 80, 82–84

liquid expansion, 6

liquid-liquid treating, 141–142

liquid phase, 5

liquid separator, 38

liquid-solid treating, 143

low dosage hydrate inhibitor (LDHI), 76

low temperature separation (LTS), 35

magnetic bearings, 126

matter, 10

membrane processes, 62–64

membranes, 146

mercaptans, 26

mercury adsorber, 95

mercury embrittlement, 92

mercury removal after-filter, 95

mercury removal costs, 94

mercury removal unit (MRU), 90

Merox process (commercial process), 142

metal sulfide on carbon or alumina, 93

metal sulfide systems, 93

methane, 8

methanol, 7

methyldiethanolamine (MDEA), 52

methyl ethylene glycol (MEG), 37

million standard cubic foot (MMscf), 100

miscibility, 3

miscible liquids, 3

mixed chemical/physical absorption processes, 60

mole, 10

molecular percent, 10

molecular sieve, 35, 93

molecular sieve process, 61, 85–86

molecular weight, 10

molecule, 2

monitoring of fractionalization plants, 139–140

MRU location, 94

multiple stage refrigeration, 46

mutually soluble liquids, 4

natural gas, 1

natural gas liquid (NGL) fractionation plants, 134-136

    about generally, 133–136

    butane splitter column, 137

    debutanizer (DeC$_4$) column, 136

    deethanizer (DeC$_2$) column, 137

    deisobutanizer (DIB) column, 137

    depropanizer (DeC$_3$) column, 137

natural gas liquid (NGL) product treating, 141–143

    liquid-liquid treating, 141–142

    liquid-solid treating, 143

natural gas liquid (NGL) recovery-cryogenic, 113–129

    about generally, 113–114

    applications, 115–122

    cryogenics, 127–129

    truboexpanders, 122–126

natural gas liquid (NGL) recovery-lean oil

  absorption, 97–111

    lean oil absorption, 98–103

    rejection system, 104–107

    separation system, 108–111

natural gas liquid (NGL) recovery operations, 77

natural gas liquids (NGLs), 35

"Natural Gasoline Specifications and Test Methods"

  (GPA Publication 3132), 138

net heating value, 12

nitrogen rejection units (NRUs)

    cryogenic NRU processes, 146–148

    NRU processes, 149–152

    process selection, 145–146

nonassociated gas, 1

nonregenerative acid gas removal, 51

off-gases, 12

oil-film resonance, 125

oil purification, 109–110

oil-to-gas ratio, 100

oil whip, 125

oil whirl, 125

operating problems, 140. See also problems

overhead product, 134

oxidation, 57

packed columns, 134

paraffins, 88

physical absorption processes

    propylene carbonate process, 59–60

    Rectisol® process (commercial product), 60

    Selexol® (commercial product), 59

physical properties comparison, 8–9

pigging, 30

pigging frequency, 28

pig receivers, 25, 28

pipeline grid, 35

plate fin exchangers, 127

poor outlet water dew point, 88–89

potassium carbonate, 51

pounds per square inch (psi), 3

pounds per square inch absolute (psia), 3

pounds per square inch gauge (psig), 3

prefractionator, 150

presaturation, 100

presaturation systems, 100

pressure, 3

pressure drop, 89, 94

pressure indicators (PIs), 44

pressure swing adsorption, 145

pretreatment, 147–148

problems

    with dew-point control and refrigeration systems, 44–46

    with ethane/butane in propane, 44

    with fractionators, 140

    in glycol injection systems, 78

    with inadequate inlet separation, 65

    with lean oil absorption, 102–103

    with methanol, 77

    in the recovery system, 102

    with rejection system, 107

    with retrograde condensation, 25

    with separation system, 110–111

    solved by the gas feed reception and condensate

      stabilization unit equipment, 33

    with sulfur-impregnated carbon systems, 92

process descriptions

    comparison of dew-point processes, 40

    dew-point control and refrigeration systems, 35–46

    glycol/J-T valve cooling process, 38–40

    glycol/propane system, 37–38

    silica gel process, 36–37

    unit specifications, 40

process engineers, 2

process selection

    cryogenic absorption, 145

    cryogenic distillation, 146

    membranes, 146

nitrogen rejection units (NRUs), 145–146
    pressure swing adsorption, 145
product specifications, 27, 138
propane, 3
propane-recovery process, 120
propylene carbonate process, 59–60

quench column, 70

Rankine scale, 2
reboilers, 16
reciprocating compressors, 32
reclaimers, 53
recompression, 148
reconcentrators, 38
recovery systems, 98–103
rectification section, 21
rectifying, 21
Rectisol® process (commercial product), 58–60
reflux, 20, 108
refrigeration systems, 35, 41
regeneration, 26
regeneration gas, 36
regenerative acid gas removal, 51
regenerative molecular sieve, 93
Reid Vapor Pressure (RVP) requirement, 28, 133
rejection system, 98
    hot rich-oil flash tank, 104
    possible problems with, 107
    rich-oil flash demethanizer, 105–106
retrograde condensation, 25
rich glycol, 77
rich oil, 99
rich-oil demethanizer (ROD), 105
rich-oil demethanizer bottoms, 105
rich-oil flash demethanizer, 105–106
rich-oil fractionator, 107
royalty charges, 59

safety instrumental systems (SIS), 28
SCOT process (commercial product), 67, 70
Selexol (commercial product), 59
sensible heat, 11
separation system, 98
    oil purification, 109–110
    possible problems with, 110–111
    still, 108–109
separation tower, 18
shell-and-tube exchangers, 16

Shell Claus Off-Gas Treating (SCOT) process, 67, 70
silica gel process, 35, 36–37
silica gels and activated alumina, 85
slipping (rejection), 54
slug catchers, 25, 30
solid desiccants, 84–85
solubility, 4
"Specification and Test Methods for LPGas"
    (GPA Publication 2140), 138
specific gravity, 3
split-vapor process, 119
stabilization column, 32
stabilizer overhead compressors, 32
stack loss, 12
staged separation, 42
standard cubic foot (scf), 12
sterically hindered amine-based solvents, 54
stills, 107, 108–109
stripping section, 20
subambient temperature, 58
Sulfinol® process (commercial product), 60
sulfur-impregnated activated carbon, 92–93
sulfur recovery, 67–70
    catalytic recovery, 68, 70
    catalytic step, 67
    thermal and catalytic reaction, 67
    thermal process, 67
    thermal reaction, 67
    thermal step, 67
sulfur recovery and Claus off-gas treating
    Claus off-gas treating, 70
    sulfur recovery, 67–70
sulfur recovery units (SRUs), 54, 67
superficial velocity and residence time, 95
sweetening acid gases, 51
switching valves for adsorption and regeneration
    operation, 90

tail gas, 54
technology selection
    design basis and specifications for dehydration
      and mercury removal, 92
    metal sulfide on carbon or alumina, 93
    regenerative molecular sieve, 93
sulfur-impregnated activated carbon, 92–93
temperature, 2–3
temperature differential, 12
temperature indicators (TIs), 44
thermal and catalytic reaction, 67
thermal process, 67
thermal reaction, 67

thermal step, 67

thermocouples, 88

thermodynamic inhibitors, 76

thiobacillus bacteria, 57–58

total installation cost (TIC), 93

towers, 67

tower with trays, 18

trays, 18

treating and processing, 23–26

triethylene glycol (TEG), 37

triple-column cycle, 149

truboexpanders, 122–126

turboexpander process, 115–119

turboexpansion, 97

turndown ratio, 26

Twister™ Supersonic separator (commercial product), 40

unit specifications, 40

upstream, 25

vacuum, 59

valve trays, 131

vaporization, 53

vaporizing, 5

vapor pressure, 5–6

vapors, 4

volatility, 14

volume expansion, 6

water dew-point control, 35

weirs, 131

wet stills, 108

working fluid, 35

Y-grade, 97